Neutron Stars

Neutron Stars

The Quest to Understand the Zombies of the Cosmos

Katia Moskvitch

Harvard University Press
Cambridge, Massachusetts
London, England
2020

Copyright © 2020 by the President and Fellows of Harvard College
All rights reserved
Printed in the United States of America

First printing

Library of Congress Cataloging-in-Publication Data

Names: Moskvitch, Katia, author.
Title: Neutron stars : the quest to understand the zombies of the cosmos / Katia Moskvitch.
Description: Cambridge, Massachusetts : Harvard University Press, 2020. | Includes bibliographical references and index.
Identifiers: LCCN 2020006175 | ISBN 9780674919358 (cloth)
Subjects: LCSH: Neutron stars. | Pulsars.
Classification: LCC QB843.N4 M67 2020 | DDC 523.8/874—dc23
LC record available at https://lccn.loc.gov/2020006175

To my sons, Tima and Kai

In memory of Fritz Zwicky

Contents

Prologue *1*

1 ✦ A Collision That Shook the Cosmos *5*
    Deeper Dive: The Origin of Gold *31*
    Deeper Dive: Why Was the Kilonova Blue? *33*

2 ✦ Discovering Neutron Stars...and Little Green Men? *35*
    Deeper Dive: The Interstellar Medium, Home of Neutron Stars *56*

3 ✦ When Stars Go Boom *60*
    Deeper Dive: Pulsar Kicks *84*
    Deeper Dive: The Death of Massive Stars *87*

4 ✦ Zombies and Starquakes *90*
    Deeper Dive: The Multibeam *126*
    Deeper Dive: The Exotic World of X-ray Sources *128*
    Deeper Dive: Pulsar Timing *129*

5 ✦ Journey to the Center of a Neutron Star   *131*

6 ✦ How Neutron Stars Keep Spoiling Dark Matter Theories   *163*

7 ✦ When Pulsars Have Planets   *185*

8 ✦ Giant Scientific Tools of the Universe   *198*
    Deeper Dive: Kepler's Laws and Beyond   *228*

9 ✦ Fast Radio Bursts, an Unfinished Chapter   *231*

    Epilogue   *255*

Notes   259

Acknowledgments   279

Index   283

Color photos follow page 134

Prologue

"I'M GOING to tell you something that's going to blow your socks off. It will be the biggest story of your career." Matthew Bailes, an astronomer from Swinburne University of Technology in Australia, is offering me a ride to my Airbnb. It's a warm evening, September 6, 2017, and we've come to the end of a long conference at Jodrell Bank Observatory, in a building next to the magnificent Lovell Telescope near Manchester, England. The event marked the fiftieth anniversary of the discovery of radio pulsars—objects far away in space that rotate rapidly, sending out powerful beams of radiation. We detect them by recording the pulses of their radio waves as they arrive at our telescopes, as well as their x-rays or gamma rays. Pulsars are neutron stars—the small but ultra-dense and incredibly magnetic objects born out of the remnants of massive stars, several times more massive than our Sun, after they have exploded in a spectacular supernova.

That the conference is being held here in England is especially significant. After all, pulsars were discovered in England in 1967, by a University of Cambridge graduate student—a young Irish woman named

Jocelyn Bell. I look up at the telescope. It's taken on a reddish hue as it basks in the rapidly setting sun, towering over us like a giant psychedelic mushroom.

Bailes knows I'm a science journalist looking for stories. I'm always looking for stories. But this time, it's different. He wants to talk—so very much. Yet he can't. Something big has just happened, something really big; there have been rumors on Twitter that a collision of two neutron stars has been detected. If true, it would be the first time that scientists and their instruments managed to catch them in the act and potentially observe more than just their pulses. The news is not public yet and won't be for a few more weeks. But if the rumors turn out to be true, the discovery is likely to answer numerous scientific questions about these enigmatic objects we barely know. It could also lift the fog on many mysteries, like the origin of brief but ultra-energetic gamma-ray bursts, and even help to confirm Albert Einstein's theory of general relativity.

But for the five thousand or so researchers in the know, keeping the news under wraps is a challenge. All day between the official talks at the conference, I could feel the buzz as scientists discussed the potential discovery in hushed tones. The news was hanging like a storm cloud over the attendees, ready to burst at any moment.

I spend the next few hours of that evening in a cryptic and at times frankly bizarre conversation with Bailes, down in a village pub a few minutes' drive away from Jodrell Bank. Inside, the smell of beer and raw yeast is so overpowering, it's probably encrusted in the walls. Bailes is very careful, hinting that something is about to break that will change astrophysical history, but without giving anything away. Like all other scientists, he is mindful of the embargo imposed by LIGO (the Laser Interferometer Gravitational Wave Observatory), which operates the detectors that first spotted the event. And I am using all my journalistic wit to get him to say more. In the end, I must leave it at a lot of "ifs" and "mights" about "possible" and "alleged" events. Even so, not long after saying goodbye to Bailes, I'm on the phone to my editor at *Quanta Magazine,* Michael Moyer.

"Something big has happened, Michael. You know those rumors on Twitter about a possible neutron star merger? It looks as if they are all true. We have to get on the story, now."

Indeed. This merger—a cataclysmic collision of two extremely dense, massive, but tiny objects in deep space—has unleashed a bonanza of insights into so many cosmic enigmas, every one of which would count as a great scientific advance. We now know where most of the heavy elements such as gold and platinum come from. But the much bigger reward has been the dawn of a whole new way of observing the Universe, called multi-messenger astronomy, and with it, the renewed attention granted to a class of peculiar objects: neutron stars.[1]

MY AIM IS to introduce you to these fascinating and enigmatic objects, as well as to the people and places involved in unlocking the mind-blowing scientific mysteries related to them. Just think of a sphere, merely twenty kilometers across, with a mass several times that of our Sun. It's spinning six hundred turns a second so regularly that in the near future it may serve as a galaxy-scale navigation system, helping to guide humans to other worlds. Neutron stars have often been overshadowed by the popular black hole; it is beyond time to take them out of the dusty astrophysics drawer.

To write this book, I traveled around the world, to the far-flung places where radio telescopes big and small listen to the Universe. You'll get there too—to the desolate, Mars-like landscape of the Atacama Desert in Chile; the rainforest of Puerto Rico; the semi-arid plains of the Karoo in South Africa; the land of kangaroos, venomous snakes, and snowy white cockatoos in remote Australia; the marshes of the Netherlands; the rain-sodden countryside of northwest England; the fields near the city of Pisa in Italy; the huge radio-quiet zone in Pocahontas County in the mountains of West Virginia; and the picturesque Okanagan Valley wine region of British Columbia.

You will also travel beyond the observatories, through space and time, to the outskirts of our galaxy and farther, into the intergalactic vastness.

You will learn how neutron stars are born from mega explosions called supernovas, which occur when a star several times the size of our Sun dies. You'll find out what happens when two neutron stars collide in a cosmic catastrophe, emitting such strong gravitational waves that we can detect them on Earth, and producing the brightest light humans have ever observed—so that each such burst, at least for a brief moment, outshines in gamma rays the rest of the entire Universe visible from Earth.

Through the work of the amazing researchers you will meet in the following chapters, you'll find out how the neutron stars we call pulsars emit radio waves as they spin—and how we can detect these waves. You'll learn how scientists observe neutron stars inside supernova remnants—or the exploded remains of a supergiant, which were known in earlier times as "guest stars." You'll encounter "millisecond pulsars," which spin as fast as a thousand times per second, yet are held so strongly together that they remain intact; "magnetars," which possess the most powerful magnetic fields known in nature; and radio pulsars that occasionally experience sudden changes in the rate at which they spin, or "glitches," that can help astronomers study strange physical phenomena inside the neutron star.

Finally, you'll learn about the recent discovery of fast radio bursts, brief pulses that astronomers are still struggling to explain but that may well be generated by neutron stars. Many radio telescopes are working tirelessly to find out more about these mysterious flashes in deep space. But even when we get the answer, the book on neutron stars won't be complete. There's so much more to uncover, in our galaxy and beyond. So whenever you go outside, no matter if you're in a city full of lights or in a faraway desert, don't forget to look up. You won't see neutron stars with the naked eye, but you'll know that they are there, spinning away, sending radio waves and ripples through spacetime. Life is so much more than what we can see around us, and that's the beauty of it.

Look up.

+ 1 +

A Collision That Shook the Cosmos

IT WAS THE MORNING of August 17, 2017, and Italian astronomer Marica Branchesi was tired. Branchesi, an assistant professor at the Gran Sasso Science Institute, had spent the night in a hospital room in the picturesque walled town of Urbino, in central Italy, comforting her younger sister Marilisa through what seemed like an endless labor. After a healthy baby Noah arrived and the family had some precious time to celebrate, it was time to go. New auntie Marica kissed her sister goodbye, smiled at Noah, and went home to get some rest.

Because it was the summer holidays, Urbino was eerily quiet. Branchesi's father picked her up at the hospital and took her home, weaving through white brick buildings that were radiating heat, making it feel even hotter than it was. Finally, the forty-year-old arrived at her family's small house, with its shady, leafy garden. It was her partner Jan's birthday and he was just starting to make lunch; Diego, their two-year-old son, and his eight-month-old brother, Damian, played quietly outside.

Branchesi fired up her laptop. She was exhausted but wanted to squeeze in some work. As a member of a global team of scientists working with LIGO, the US-based Laser Interferometer Gravitational-Wave Observatory, and Virgo, a gravitational wave detector near Pisa, Italy, she knew that just three days earlier the instruments had spotted a signal from a collision of two faraway black holes.[1]

It wasn't the first such discovery; LIGO had logged five similar events since the most recent upgrade of its instruments. It had detected the very first black hole merger ever observed on September 14, 2015, just days after it was switched back on. With LIGO, it had finally become possible to directly spot gravitational waves: ripples in spacetime, triggered by cataclysmic collisions between extremely dense objects far away in space that wash over the Earth much like ripples from a stone hitting the surface of a calm pond. The observation had proven Albert Einstein right; a century earlier, he had predicted that gravity could create such waves and send them across the Universe at the speed of light. On October 3, 2017, three of the four lead researchers behind LIGO—Kip Thorne, Rainer Weiss, and Barry C. Barish—would bag the Nobel Prize in Physics for making the detection of black hole mergers possible.[2] (Ron Drever, who was also key in devising the technology for LIGO, passed away just half a year before this Nobel Prize was announced.)

Branchesi, however, was a little disappointed in the news. While she was excited about this most recent black hole merger, she was really hoping for LIGO to spot something else: ripples from a collision of two objects even more mysterious than black holes. She was hunting for waves from a merger of two neutron stars—the small, ultra-dense, and rapidly spinning leftover cores of massive stars that had run out of fuel and collapsed under their own gravity. Unfortunately, LIGO was just one week away from being shut down for two years for a planned upgrade, and so far it had detected black hole mergers only. Good, but for Branchesi not good enough.

"Food is ready!" Jan called from the garden, where he had set up the table. "Food, Mummy!" Diego excitedly repeated after him. Branchesi closed the laptop and joined her family for lunch. After finishing the salad, she picked up Damian and took Diego by the hand, leading them to their bedroom for a nap. Exhausted after the sleepless night, she was also hoping to finally get some rest. Just then, her phone chirped.

Branchesi didn't get a chance to rest that day. In fact, she didn't get much sleep over the next ten days.[3] The chirp was an email alert, asking her to join an urgent teleconference organized by colleagues in the LIGO collaboration. Reading the email made her gasp. The two LIGO detectors, together with Virgo, had just spotted what she had been waiting for: the unmistakable signal of two neutron stars crashing into each other, some 130 million light-years away. Although neutron stars are less dense than black holes, their collision had been violent enough to release huge amounts of energy in all directions, sending ripples through spacetime.

And now, these ripples had finally reached Earth.[4] Branchesi closed her eyes for a moment. Had LIGO not been upgraded two years earlier, the ripples—incredibly weakened by their 130-million-year-long journey—would have come and gone completely unnoticed. Scientists would have been none the wiser. This time though, thanks to Branchesi and a handful of others, astrophysicists, astronomers, and gravitational wave physicists had been prepared for the unlikely encounter.

She looked at her little boys. Diego had already heard plenty of stories about black holes from his parents; soon his mom would be adding a few about neutron stars. Branchesi knew that—if confirmed—this merger, now known as GW170817, could well be the defining moment of her career. It could be the culmination of her decade-long effort to bring together researchers from many different fields to collaborate for a new approach: multi-messenger astronomy.

A "messenger" is any kind of signal coming from space. Our Sun, for example, emits not just light, but also a steady stream of nearly massless particles called neutrinos.[5] So far, multi-messenger astronomy has been used to confirm three events occurring beyond our Milky Way. One was the detection in 1987 of a supernova using optical telescopes and observatories that captured neutrinos. Another one was the pinpointing in 2018 of the origin of a cosmic neutrino to a blazar—home of a supermassive

Marica Branchesi
(Courtesy of Marica Branchesi)

black hole at the center of a galaxy four billion light-years away—using optical telescopes and a neutrino detector in Antarctica called IceCube.[6] The game changer, however, would turn out to be the project that Branchesi had been working so hard on: the detection and observation of GW170817. This time, gravitational-wave physicists received a signal of ripples in spacetime washing over Earth and immediately alerted astronomers, so that they could point their optical, radio, and all other telescopes designed to capture various wavelengths across the electromagnetic spectrum toward the source of the ripples, in an effort to observe the full range of signals—or messengers—from the cataclysmic event that had triggered the wave.

Crucially, this historic collision and the nascent field of multi-messenger astronomy has helped scientists to begin understanding the composition and evolution of one of our Universe's mind-bendingly odd objects: neutron stars.[7]

By the following year, Branchesi's work would propel her onto *Time* magazine's list of the one hundred most influential people of 2018.[8] But even before the results were in, on that hot afternoon of August 17, 2017, standing in her garden in Urbino, she knew that this event would be forever etched in future science and history books.

Eleven Billion Years before the Detection

Look up at the Moon one night, when it's full. Then imagine putting a dot on it with a pen. The Moon is 3,476 km (2,159 miles) across; the dot would be less than 1 percent of that—some 20 km. That's a tad smaller than the diameter of Chicago if it were curled up as a ball floating in space—and that's the size of an average neutron star.

A neutron star is what's left of a once-massive star that originally had eight to fifteen times more mass than our Sun. Having burned through its nuclear fuel for millions of years, the parent star eventually runs out of fuel and dies in a spectacular supernova explosion, a stellar death that in a galaxy the size of our Milky Way is expected to happen on average once every half a century. Compared to the vast emptiness of the Universe, a neutron star may appear to be a rather unremarkable object, were it not for its epic density—a hundred trillion times the density of water. A dense object has a lot of matter shoved into a very small space; neutron stars are the densest objects we know of that are made of normal matter. Add just a bit more matter to a massive neutron star, or merge two of them into one, and it will collapse into a black hole. Our Sun is 1.4 million km across, but it has nearly the same mass as a tiny neutron star merely 20 km across. If you tried to scoop up some neutron star matter with your finger, like a child stealing some cookie dough, even the tiniest amount would pull you down with the weight of one billion tons.[9]

As if this level of density wasn't enough, consider that these weird zombie stars also zoom through space at high speed, while rapidly spinning around their own axis once every second or even faster. Some neutron stars also emit jets of electromagnetic radiation at their magnetic poles. As a neutron star rotates, when one of its jets happens to point in the direction of Earth, we can detect the radiation as pulses of radio waves. A spinning neutron star, then, is a bit like a lighthouse, which, as it revolves, shines with a continuous stream of light, but appears as a flash to ships at sea, or in our case, astronomers on Earth. Such neutron stars are called pulsars, and they are the ones we typically detect. So regular are the pulses of many of them that there has been a recent push to use pulsars to cross-check the atomic clocks that define International Atomic Time.[10]

Astronomers believe that there are as many as a hundred million neutron stars in our galaxy alone, although so far we have detected only close to three thousand radio pulsars.[11] And yet, we know very little about them.

We knew even less before 2:41 p.m. local time on August 17, 2017, when Marica Branchesi received the email alerting her to the neutron star merger in a small galaxy nearby.[12] Finally, humanity would have the chance to learn much more about these strange objects.

The two neutron stars spotted by LIGO and Virgo were probably created about eleven billion years ago, when the Universe was young and neither Earth nor our Solar System existed. Back then, ordinary stars were busy grouping themselves into clusters. Two of these stars, each perhaps ten times more massive than our Sun, died—one after the other. They were not too far apart in cosmic terms, and their leftover cores, each now with a mass ever so slightly more than one solar mass (the mass of the Sun), began spiraling toward each other, pulled by mutual gravitational attraction. It was a fateful dance. Orbiting their common center of mass, they tugged on the fabric of space and time, warping it—similar to how a rolling bowling ball dents a bedsheet that's held stretched out at its four corners. The warping caused by these neutron stars caused ripples in spacetime—gravitational waves that propagated through the Universe.[13]

130 Million Years before the Detection

While the two neutron stars spiraled toward each other, the Universe was evolving and expanding, forming galaxies and giving birth to billions of new stars. Then, 130 million years ago, they got so close to each other that they created tides on each other's surfaces, a bit like when the Moon pulls on Earth's oceans. These tidal effects disrupted the two stars, stretching and squeezing them.

Not much later, the neutron stars finally collided, in a cosmic cataclysm. Part of the ejected material stayed bound to what remained of the pair, creating a disk of debris all around the remnant object—a so-called accretion disk—that in turn generated a jet of material that rushed through the Universe at nearly the speed of light, emitting radiation in x-ray, optical, and radio waves. The jet also produced a brief burst of ultra-energetic gamma rays, the most energetic electromagnetic radiation we know.

Some of the ejected mass managed to get away from the gravitational pull of the remnant of the collision, creating an extremely hot and rapidly expanding cloud of debris shaped like an expanding doughnut. This cloud was so neutron-rich that it triggered the formation of elements heavier than iron, such as gold, silver, and platinum. Astronomers estimate that there must have been heavy elements some ten thousand times the mass of the Earth in that cloud, creating 236 sextillion (that's twenty-one zeros) tons of pure gold alone—an amount equal to forty times the Earth's mass. The radioactive decay of all these heavy elements generated light—an optical emission from the radioactive afterglow called a kilonova.[14]

As the two neutron stars merged, the density of the new body surged further. In all likelihood, the combined neutron star became too massive to survive, collapsed in on itself, and formed a black hole. Crucially, the merger greatly amplified the gravitational ripples from before the collision. These ripples became the invisible messengers of the almighty clash, scurrying in all directions at the speed of light with the energy of two hundred million Suns.

When all this happened, on Earth the dinosaurs of the early Cretaceous were roaming the lands and waters. The ripples did not reach our world and tickle LIGO's and Virgo's sensitive tools until August 2017. During those 130 million years, the gravitational waves became significantly weaker, moving—as predicted by Albert Einstein's general theory of relativity—at the speed of light, or 300,000 km per second. Other signals from this cosmic event were also on their way to us: light and radio waves, traveling at the same speed. As with everything we see in space, we are observing the past. Even the light from the Sun takes eight minutes and twenty seconds to reach us, so if the Sun were to suddenly vanish (not that it will in our lifetimes), it would take exactly that long before we would know it.

When LIGO and Virgo spotted the waves, they sent out automated alerts. These alerts were received by a few so-called first respondents, whose job it is to look at all potential candidates flagged by the software. The data shared by the alerts were clear and dramatic—the strength of the signal meant that the two objects were in the expected mass range of neutron stars, in other words smaller than black holes, which meant—in theory—that the collision should also give off light. Sure enough, just two seconds after LIGO and Virgo caught the gravitational wave signal, the Fermi Gamma-ray Space Telescope also detected a bright flash of gamma radiation. Within minutes, the first respondents notified the wider LIGO-Virgo collaboration, including Branchesi, and kicked off what would be a very long and historic teleconference.[15]

A Century before the Detection

Remember that story about Isaac Newton getting hit on the head by an apple and suddenly figuring out how gravity worked? That's not exactly how his epiphany happened, but he did observe apples fall in his garden at Woolsthorpe Manor near Grantham in Lincolnshire, England—from a specific tree, which apparently is still growing in that garden, three centuries later. Newton was puzzled by the fact that the apples always fell to the ground, and that prompted him to develop his theory of universal gravitation, which he published in 1687. The theory describes gravity as a force that acts on all matter in the Universe and depends on both mass and distance. According to this theory, each and every particle of matter attracts any other particle with a force directly proportional to the product of their masses and inversely proportional to the square of the distance between them.[16] And so the law stood, unchanged and unchallenged, until Albert Einstein came along a couple of centuries later.

For Einstein, a clerk in a Swiss patent office who toyed with physics in his spare time, gravity was not a force at all. He argued that it is instead a distortion of space and time, or spacetime, which includes four interlinked dimensions: the three dimensions of space (up-down, left-right, forward-back) plus the one dimension of time. Published in 1916, Einstein's general theory of relativity states that what we perceive as the force of gravity actually comes from the curvature of spacetime. Massive objects such as stars and planets make the spacetime fabric bend and warp, creating mountains, wells, and ridges that force nearby objects moving through spacetime to zigzag or go up and down. While gravity seems to pull the Earth toward the Sun, making our planet rotate around it, it's simply the curving of spacetime around the Sun, which is much more massive than the Earth, that is dictating how our planet moves.

Einstein also showed, mathematically, that any accelerating mass that isn't perfectly spherically symmetric would distort spacetime and trigger gravitational ripples that would travel at the speed of light across the Universe. Even waving your hand around generates gravitational waves, but they are too small to be detected; to really distort spacetime in a measurable way requires an immense amount of energy. Think instead of catastrophic cosmic events involving massive accelerating objects, such as black holes and neutron stars orbiting each other and then colliding at a third of the speed of light. A collision like that, according to Einstein, would trigger incredibly energetic gravitational waves that would wash over planets, stars, and everything else, carrying with them information about their source and perhaps even about the nature of gravity. Einstein came back to these ripples in spacetime several times in his later work, but for decades their existence remained purely theoretical.[17]

In 1974, astronomers Russell Alan Hulse and Joseph Hooton Taylor Jr. of the University of Massachusetts Amherst indirectly proved the existence of gravitational waves. They noticed that in a system of two gravitationally bound neutron stars, the orbital period—how long it takes them to go around their common center of mass—gradually shrinks. The two bodies move closer and closer to each other on an inevitable collision course, because, Hulse and Taylor figured, the system loses energy in the form of gravitational waves.[18] The neutron stars the pair studied will collide in some three hundred million years. The system is now called the Hulse-Taylor pulsar, and in 1993, the astronomers received the Nobel Prize in Physics for their work.[19]

But Hulse and Taylor's observation was not direct evidence. To detect gravitational waves directly, scientists needed new and extremely finely tuned instruments. The result: the twin LIGO detectors—in Hanford, Washington, and Livingston, Louisiana—which work together, using the interference of two beams of laser light to measure distance very, very precisely. The observatory is run by the LIGO Scientific Collaboration, a group of more than a thousand scientists at universities in sixteen countries around the world. Operated by MIT and Caltech, LIGO was first proposed in the 1980s by Rainer Weiss, Kip Thorne, and Ronald Drever, but because major scientific projects require substantial funding and need to jump numerous bureaucratic hurdles, it took another decade before construction began. LIGO finally came online in 2002.[20]

Five years later, in 2007, it was joined by a third detector, the EU-funded Virgo near the Italian city of Pisa.[21] Because it took the help of Virgo for scientists to triangulate the exact location of the first observable neutron star merger, I decide to visit it as part of my quest to get closer to the amazing tools, the results of human ingenuity, that are slowly but tirelessly peeling back the layers to reveal more about our most compelling cosmic mysteries.

I START IN PISA, where I ascend the leaning yet never quite falling tower, but don't linger: the four-degree tilt plays tricks on my brain, and I quickly start feeling slightly motion sick. The tower is also arguably the best-known location of a gravity test—albeit one that probably never happened. Historians agree that the Italian scientist Galileo Galilei's dropping of two spheres of different masses from the tower was merely a thought experiment. Still, I instinctively look out for objects hurtling toward the ground—who knows if someone might try to repeat Galileo's alleged feat. To prevent this from happening, all visitors are now obliged to store any bags in a free locker room before going up the tower.

But it's not really the tower I'm here to see. About twenty minutes' drive from Pisa, in the beautiful Tuscan countryside, I find the village of Santo Stefano a Macerata. Nearby is its unusual scientific attraction: two low, demi-cylindrical tunnels, each three kilometers long. They are of an unassuming, somewhat blue, somewhat light-purple color that makes them blend into the environment. "Periwinkle," I hear another visitor looking at the tunnels say, and he is spot-on. The tunnels are positioned at a right angle to each other and nestled in fields; they stretch all the way to a mountain range in the distance, going under several tiny bridges. A drone flying above would see them form a giant letter L. This is the Virgo interferometer, named after the Virgo Cluster of about 1,500 galaxies in the Virgo constellation, some 50 million light-years from Earth.

The construction of the detector started in 1996; it is managed by the European Gravitational Observatory (EGO) and run by a collaboration of scientists from France, Italy, the Netherlands, Poland, and Hungary, although France and Italy are responsible for nearly half the operational cost. The location is actually not the best, says experimental physicist Valerio Boschi, who is accompanying me on the tour. It's too close to the Mediterranean Sea and to the city of Pisa—both of which are sources of seismic disturbances that create interference. Minimizing this noise is crucial to detect the ultra-weak signals from gravitational waves traveling to us from far away in space. But when Virgo was conceived, it was more important to find a large enough plot of land to build arms three kilometers long. The founding organizations, the French Centre National de la Recherche Scientifique (CNRS) and the Italian Istituto Nazionale di Fisica Nucleare (INFN), had to negotiate with many farmers to buy the land needed to build the detector.[22]

The work was completed in 2003 with a final cost of around 250 million euros (by comparison, the cost of LIGO's twin detectors was about 600 million euros). And in the last few years, after the detection of gravitational waves, the number of visitors to Virgo wanting to have a look at the facility has skyrocketed. "We had about a thousand in 2015," Stavros Katsanevas, the EGO director, tells me. "In 2016, it was two thousand, and this year, 2019, we had eight thousand visitors. We are finding it more and more difficult to cope!"[23] Indeed, on the day I'm here, I see three groups who arrived on tourist buses parked nearby: two groups of schoolchildren and another of university students.

After a quick look around the visitor center, and a short walk with plenty of lizards scurrying through the grass, I enter one of the two demicylindrical tunnels housing a long, shielded pipe 120 cm in diameter. An identical pipe runs in the other tunnel, and inside both pipes the instruments operate in an ultra-high vacuum. Both Virgo and LIGO work very similarly, trying to detect waves at frequencies between 10 Hz and 10,000 Hz (10 kHz). At the joint of the pipes is a small building where I am asked to walk very carefully so that I don't shake the floor too much during data taking. There, a laser emits a beam that is split in two. Each half of the beam is sent to the far end of each perpendicular arm at exactly the same time, where they hit a mirror. The mirrors at Virgo are suspended by a "superattenuator" (a prototype is displayed at the entrance of the main building)—an extremely precise mechanism that limits as much as possible any disturbances that could shake the mirror, such as an earthquake or a tractor going past. It's a fairly simple system of suspended weights, one underneath the other, with the mirror at the very bottom.

"It's like an experiment we ask kids to perform," Boschi tells me. "You fill a bottle with water and suspend it, then put another bottle underneath, and a few more in the same manner, then if you sway the whole thing, the very last bottle close to the ground won't spill any water even if you sway the top quite a lot."[24] At LIGO, the system is slightly different and with more active elements than here at Virgo.

After hitting the mirror, the laser light is reflected back—and this is repeated four hundred times, reflecting the beams back and forth so that the instrument's optical length is extended from three km to 1,200 km. When the two beams are finally recombined, researchers analyze the result. If there are no gravitational waves, the two beams should return to

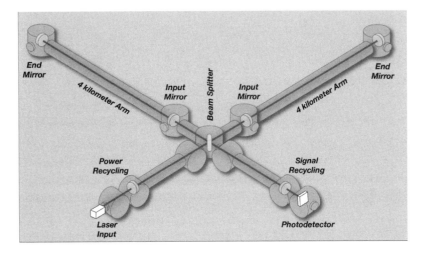

A gravitational wave moving orthogonally to the detector plane will lengthen one arm and shorten the other. LIGO, Virgo (*pictured*), and other ground-based detectors can register waves longer than the detectors' lengths—three to four km (1.9 to 2.5 miles). Such waves correspond to periods of a few hundredths to a few thousandths of a second. (Caltech / MIT / LIGO Laboratory)

the joint at exactly the same time, with no interference at all. Should a gravitational wave pass through the detector, however, local spacetime would be disturbed on very tiny scales that humans are unable to feel. The LIGO twins and Virgo can sense only very high-frequency waves, proportionate to the length of their arms.

When a gravitational ripple arrives, it makes one of the two tunnels ever so slightly longer or shorter—by less than one-ten-thousandth of the diameter of a proton, a subatomic particle. When one arm gets shorter, the other one gets longer, and then they switch roles. The amount by which the arms expand or contract depends on the amplitude of the wave—the maximum amount of displacement of a particle from its rest position, or the distance from that rest position to the crest of the wave. The change in the length of the paths makes the reflected light travel slightly out of phase with the light that was sent out: in other words, the two beams arrive back at their origin at slightly different times.[25]

Researchers then carefully measure the amplitude and frequency of the phase shift, which enables them to study the properties of the gravitational wave. When they analyze the signal alongside the findings from another gravitational wave detector, they can determine the location of the source—the faraway cosmic cataclysm that sent out these ripples across space and time. For that, the more detectors around the world the better. Today, triangulating the location of the source of any detected wave involves coordinating Virgo and the two LIGO instruments. Soon similar detectors in Japan and India will give them a hand, making the pinpointing of sources much more accurate.

Next-generation tools such as the Einstein Telescope and Cosmic Explorer, if built, will be much more sensitive than LIGO and Virgo. They are both in the proposal stage, but the plan is to have 10 km-long, above-ground arms for the Einstein Telescope, and whopping 40 km-long arms for Cosmic Explorer. These detectors will be so sensitive and precise that they will be able to spot the continuous emission of gravitational waves from neutron stars set on a collision course, but still millions of years away from merging to become a pulsar such as the one observed indirectly by Hulse and Taylor.

Seventeen Years before the Detection

For nearly a decade after their launch, LIGO and Virgo produced no results. Scientists worried from the start that this might happen, and major upgrades had always been planned. In 2010, both LIGO and Virgo were shut down. Five years and $620 million later, Advanced LIGO came online. Tests started in February 2015, and the official start of the first observing run happened on September 18, 2015, with the two LIGO facilities conducting their science observations at about four times the sensitivity of the original LIGO.

Virgo was first upgraded in 2011, with its sensitivity getting a ten-fold boost. And then it shut down again, to get another major upgrade that would turn it into Advanced Virgo—which started taking data in August of 2017. Until then, the two LIGO detectors were operating on their own.

The upgrade and all the previous years of effort paid off—even before the formal relaunch of (Advanced) LIGO. On September 14, 2015, while still in engineering mode, LIGO made history. It picked up a gravitational wave signal from the collision of two stellar black holes—a "binary black hole merger," in science speak—some 1.3 billion light-years away from Earth. Announced a few months later, on February 11, 2016 (and preceded by plenty of rumors on social media), the detection, gravitational waves, LIGO, and Einstein dominated the press, researchers' watercooler talks, and even family dinners. The following year, Weiss, Thorne, and Barish shared the Nobel Prize in Physics for the discovery.[26]

Marica Branchesi, however, was already looking far beyond the scientific importance of the first observed black hole merger. In 2012, after returning to Italy from her research stint in the United States, she had been given a grant for what at the time seemed to be a rather improbable project: persuading researchers working on the (then still theoretical) discovery of gravitational waves to collaborate with astronomers who were using the electromagnetic spectrum—light, x-rays, gamma rays, or any other wavelength, in short, more traditional ways of making observations. This way, the thinking went, as soon as a gravitational wave was spotted, it might be possible to observe its source with other instruments—thus capturing different faraway cosmic "messengers" or signals from the same event, in a methodology that would come to be known as multi-messenger astronomy.

At the time, however, many astronomers were doubtful that gravitational waves would ever be observed directly. "The idea of astronomers and gravitational wave researchers working together seemed peculiar to many," says Branchesi.[27] That's not to say that scientists didn't think gravitational waves existed—there was little doubt about that. They had to. The math of Einstein's equations demanded it, so the general consensus was that powerful cataclysmic events had to disrupt spacetime and trigger ripples.

But Branchesi wasn't primarily interested in the waves themselves. Rather, she was keen to find out what else could be seen of the events that caused them. And for that she needed LIGO and Virgo to catch a wave and roughly identify the location where it came from—so that telescopes observing across the electromagnetic spectrum could straight away turn in the right direction and check whether anything else might be seen. If they observed any immediate afterglow of the kind of cosmic cataclysm that was assumed to trigger gravitational waves, astronomers would have proof that gravity ripples moved at the speed of light—confirming Einstein's century-old predictions. They would also be able to study the nature of the source, its environment, and the mechanisms that power such a huge release of energy. Black holes don't give off enough light—they are mere gravitational fields that remain after the collapse of extremely massive stars, and no matter can escape them to radiate light. So black hole mergers were of not much use to optical telescopes.[28]

Enter neutron stars. In 2010, astrophysicist Brian Metzger at Columbia University suggested that a collision of these dense objects should not only disturb spacetime and trigger ripples, but also produce a kilonova, a bright flash about one hundredth of the brightness of a supernova, which should be visible using optical telescopes. The kilonova, he reckoned, would happen in a hot cloud of radioactive debris, the material thrown out during the collision and the result of the radioactive decay of heavy elements synthesized in the merger. To top it off, the collision would also generate a brief flash of gamma radiation—a short gamma-ray burst, or SGRB. Its afterglow should then be visible in radio, x-ray, and eventually even in optical light, provided astronomers knew precisely where and when to look.[29]

Getting researchers in different fields to work together is never an easy task. It was especially tricky because gravitational waves were still very theoretical, while optical observations were long-established science. "Astronomers were skeptical about seeing an electromagnetic counterpart, because they thought that detecting gravitational waves was too difficult," says Branchesi. "Many people were rather pessimistic." Researchers typically have to fight to secure time on powerful telescopes, so why bother wasting valuable observation time to hunt for the byproduct of ultra-weak waves that no one was sure could ever be detected? Branchesi remembers going to conferences and feeling totally helpless, with astronomers looking at her with blank faces—or not looking at all.[30]

It took a year of discussions and email exchanges before any formal meetings even started taking place, around 2012. Over the next decade, at various events, Branchesi—who by then had become a co-chair of the collaboration's electromagnetic follow-up group—and several of her colleagues who also believed in multi-messenger astronomy, spoke to more and more astronomers. One by one, they persuaded them to join the LIGO-Virgo collaboration, and to be ready for a possible future detection. "Marica was always saying that it's really worthwhile to invest time and effort—she was always assuring us that LIGO and Virgo would give us the information and that we could really rely on it being real and robust," says Stephen Smartt, an astronomer at Queen's University Belfast. "She was a great go-between and made sure LIGO understood what we wanted and that we understood the scientific content of the information that we were getting from LIGO."[31]

Today, more than 1,200 scientists from more than one hundred institutions and eighteen countries are LIGO members, and more than five hundred are at Virgo. The two organizations are working closely together—collaborating with more than two thousand astronomers from thirty-five institutions and eleven countries. Branchesi's persuasions have clearly paid off.

Already back in 2012, the scientists at LIGO insisted that any astronomers joining the collaboration would have to sign a memorandum of understanding. They had to agree that any detection data would first be shared internally but kept from the outside world—to give everyone enough time to complete their observations and analysis and then publish at the same time.[32] Excitement among researchers kept mounting—especially when the Advanced LIGO was finally switched on. But even after the detection of several black hole mergers, few expected the discovery of gravitational waves from a neutron star merger before 2020 or 2021, when LIGO and Virgo would be upgraded again to even higher levels of sensitivity.

Still, the scientists who had joined the multi-messenger effort were ready. When Advanced LIGO started its observation run in September 2015, about eighty groups from around the world were on standby. After just four days they were alerted to the very first discovery of a black hole merger, and they swiveled more than two hundred instruments to look in its direction. They didn't see anything, because black holes don't give off light (although some have theorized that binary black hole collisions may produce electromagnetic radiation). But that was okay; the effort showed that the astronomical community was able to coordinate a broad range of instruments to extract information about a cosmic event—bringing together the precise laser beams at LIGO and Virgo, the high-energy neutrino detectors of the IceCube Observatory deep under the ice of Antarctica, and all the telescopes designed to capture any kind of electromagnetic waves given off by any object or event. "The follow-up for the black hole merger showed the astronomical community was ready for this science," says Branchesi. Netting a signal in gravitational waves and light was just a matter of time. And so they settled in for a waiting game that most expected to last the best part of a decade.[33]

Two Weeks before the Detection

August is often a quiet month in science, for those researchers working in the Northern Hemisphere, anyway. Many of them, just like workers in the rest of the world, end up for a week or two somewhere on the beach or hiking in the mountains rather than in the lab. In August 2017, LIGO was also about to take a break for a year and a half. The detector was nearing the end of its second observing run, which had been going on since the end of the previous year.

Virgo, however, had only just been turned on after its upgrade, finally joining the two LIGO instruments on August 1. The Virgo researchers had been sprinting to complete their upgrade, to secure at least a few weeks of working alongside LIGO. Observing as a trio for the first time, on August 14 the detectors spotted a binary black hole merger—the eighth in total—and the one that Branchesi was analyzing that hot summer day in Urbino.³⁴ With just a couple of weeks to go before the end of the run on August 25, the LIGO and Virgo teams were beginning to wind down, exhausted from a hectic grind of weekly and often daily teleconferences with collaborators as well as round-the-clock shifts to spot events in real time. For months now, whenever there was an automated alert, even in the middle of the night, Branchesi had to assess the need for any follow-up observations.

By the beginning of August, Branchesi's mind was already on the future, preparing for the third observing run, which would bring together all three detectors in 2019 for the first extended period following their upgrade.³⁵ On the afternoon of August 16, astrophysicist Samaya Nissanke, Branchesi's colleague at the University of Amsterdam, gave a talk at an international workshop on gravitational waves, outlining the future of multi-messenger astronomy. By the 2020s, she told her audience, she was certain that detecting neutron star mergers would be a fairly routine affair. After Nissanke's talk, a fellow collaborator from India, Bala Iyer from the Indigo-Indian gravitational wave consortium, approached her. When would we catch the first one? Iyer wondered. Boldly, Nissanke made a bet: the interferometers would detect a neutron star merger by the end of 2019—in just over two years. The very next day, she woke up to the news that her prediction had already come true.³⁶

August 17, 2017: Detection Day

When a gravitational wave washes over Earth and a detector spots that tiny offset in the arrival times of the reflected laser beams, LIGO and Virgo's software automatically uploads a record of the event to a special database of gravitational-wave candidate events. Just after midday on August 17, 2017, LIGO's systems flagged just such an event, issuing an alert that this could be a merger of two neutron stars—with a remarkably low likelihood of being a false alarm. The two objects together had a low mass, merely 2.7 times the mass of the Sun—much lighter than a stellar black hole but right in line with the known masses of binary neutron stars (a system composed of two gravitationally bound neutron stars that will eventually merge). The software flagged it as having a 100 percent chance of being observable in the electromagnetic spectrum as well.

The system calculated that the merger had happened around 130 million light-years away, which is relatively close to Earth. That was lucky, because given LIGO's sensitivity at the time, any event happening at more than about three times that distance would not have been detectable at all. The wave arrived at Virgo at 12:41:04 Coordinated Universal Time (UTC), and 22 milliseconds later washed over the LIGO-Livingston detector in Louisiana. Fast-forward another three milliseconds, and it touched the Hanford tunnels in Washington.[37]

Physicist Cody Messick at Pennsylvania State University was the first person to get LIGO's alert that day. One of the so-called first respondents, his job was to verify the possible candidates of black hole or neutron star collisions and, if confirmed, notify the wider astronomical community. It was shortly after 8 a.m.; he had just woken up and had been planning to take the day off to recover from a neck injury. At 8:43 a.m., reaching for his phone, he saw the LIGO alert—precisely two minutes after the wave had rolled over Earth. His first reaction was annoyance, because it seemed as if only the Hanford detector had picked up the signal—typically, LIGO algorithms flag events in real time only if both detectors see them. Still, the system marked the possible merger as highly significant, piquing Messick's interest. He texted his dissertation adviser Chad Hanna, who immediately replied that there had also been an alert issued from the Fermi Gamma Ray Space Telescope. Fermi had detected a short two-second pulse of gamma radiation coming from the same location, just 1.7 seconds after the arrival of the gravitational wave. Was it an optical counterpart to a cosmic merger of two objects—in other words, an observation of the same event in the electromagnetic spectrum?

Straight away, Hanna and Messick (who was still wearing his pajamas) started a teleconference with graduate students from Penn State to check the readings for mistakes. They quickly agreed that nothing was obviously wrong with the signal and decided to send out an alert to the rest of the LIGO and Virgo collaboration. The task fell to Messick, because, he recalls, Hanna was "shaking too much from excitement to type."[38] For a brief moment, they were the only people in the world who knew what had just happened.

The alert Messick sent was the one that popped up on Branchesi's phone when she was about to lie down for a nap after her long night at the hospital supporting her sister and the new baby. Within minutes, the teleconference became crowded, as more and more members of the collaboration, including Branchesi, joined. Everybody was excited, but there was also plenty of skepticism, because the second detector in Livingston hadn't flagged the event. As it turned out, a glitch in the system had resulted in a reporting fluke; some technical error—maybe a mirror slightly jiggling—had generated noise, masking the signal from the gravitational wave. Looking at the plots with the naked eye, however, made it obvious: the wave signal was real, lasting around six minutes, while the noise from the glitch that had tricked the software to disregard the wave signal was just a few milliseconds long. Both LIGO detectors had now confirmed the data.[39]

Meanwhile, in Italy, Virgo scientists had also realized that their instrument had spotted the historic signal—a faint one, but an unmistakable signal nonetheless. "People were so excited, they kept constantly talking over each other on the telecon," says Branchesi. "It was mad."[40]

Soon, astrophysicists from all over the world were piling into the call. "Although I know I shouldn't have, I stepped away from the telecon for a minute to call my wife and my dad to tell them the exciting news," says Messick.[41] The online chat of the telecon exploded with demands to send an alert to all the optical telescopes for immediate follow-up observations, called a GCN circular (GCN is short for gamma-ray burst coordinates network). Everyone was eager to start the search for the source of the wave, across the entire electromagnetic spectrum. It had taken Branchesi and her colleagues a decade to lay the ground for multi-messenger collaboration; now, finally, all their work was about to pay off.

Just forty minutes after LIGO's and Virgo's original detection, at 13:21:42 UTC, the GCN circular alert hit the inboxes of astronomers around the world. In a matter of hours, robotic and survey telescopes began to swivel to see the aftermath of this cosmic cataclysm from the distant past.[42] The age of multi-messenger astronomy had begun.

Spotting the collision in optical light quickly turned into a race. Astronomers the world over rushed to their telescope controls. Not everyone was at the starting line, though: Ryan Foley from the University of California, Santa Cruz, was enjoying a leisurely day out with his partner in Copenhagen's Tivoli gardens, on a short break squeezed in before the end of LIGO's observation run. Suddenly his phone buzzed, alerting him to a text message from Dave Coulter, a colleague back home in California. Foley read the text and after a quick and apologetic goodbye to his partner, he immediately hopped on his bike to get to the University of Copenhagen as fast as he could. He wanted to jumpstart the follow-up observations with the one-meter Swope telescope at Las Campanas Observatory in Chile. It took five hours to combine the initial coordinates from LIGO and Virgo and obtain as precise a location as possible. Eleven hours after the arrival of the wave, Foley's team identified a bright flash in the near-infrared spectrum at the point in the sky specified by the gravitational wave detectors. The discovery prompted Foley to send a text to a colleague, Jess McIver, stating simply: "I think I found it." The team was the first to capture the optical images, winning the race—and the optical counterpart was named the Swope Supernova Survey 2017a (SSS17a).[43]

Others were not far behind. The telescopes high up on the mountains of Chile's Atacama Desert were especially busy. Right after Branchesi dropped out of LIGO's overcrowded teleconference, she jumped on another call, with her colleagues at the Gravitational Wave InafTeam, or GRAWITA—the Italian collaboration formed specifically for follow-up LIGO observations in the radio, optical, near infrared, x-ray, and gamma-ray bands of the electromagnetic spectrum.[44] The team started to remotely swivel the Rapid Eye Mount Telescope at the La Silla Observatory in the Atacama—and got an image of the collision's blast in optical light just thirteen hours after the wave arrived.

A few teams had a slower start. About an hour after Messick sent the alert, Harvard astronomer Edo Berger was struggling to stay awake during a boring faculty meeting in his office. The strong coffee from the university canteen wasn't having much of an effect. When his mobile phone rang, he put it on silent. Then his desk phone rang. His colleagues stopped talking; he had to pick up. "What's up?"[45]

Berger listened, his eyebrows climbing ever higher. He put the phone down and declared the meeting over, effectively kicking everyone out. The next few minutes Berger spent whizzing through a flurry of text messages and emails, including the very first official and automated LIGO alert with the estimated coordinates of what appeared to be a collision of two neutron stars—which, it seemed, LIGO had registered just as Berger was picking up his coffee.

Thanks to the multi-messenger efforts of Branchesi and others, Berger was ready to act—just like the scientists working at some seventy other optical telescopes around the globe. It was time to zoom in on the aftermath of the neutron stars' merger. Since Fermi had reported a gamma-ray burst, Berger knew that there was a very strong chance that the event could be observed in other wavelengths. He needed his team, and fast.

Kate Alexander, Berger's PhD student at the time, was just waking up in her apartment in Boston. It was her last year of graduate school, and she was leading the radio observation efforts for Berger's team of astronomers. Still in bed, she scrolled through university emails and saw the LIGO alert. Then she saw Berger's email with a subject line: "Meet me in my office in five min!" A three-minute shower was followed by a sprint to the campus—and by 10 a.m., just two hours after the merger detection, she was barging into Berger's office.

Together with a few colleagues, they hatched a plan: they would use an optical telescope to determine the location and study the source—looking for the radioactive glow of the kilonova, the ejected debris cloud. It would show the heavy elements that should have been created and then ejected by the force of the collision, and so should show up across the electromagnetic spectrum. "Everybody was very excited and trying not to let excitement get in the way of what needed to be done," said Alexander.[46]

For any observations in the electromagnetic spectrum—whether you're hoping to document visible light, radio signals, or any other wavelength—it's important to know at which part of the sky you should point your telescope. Luckily, with Virgo's help, it was possible to triangulate the signal. Without Virgo, the "error box"—the probable area of origin for the signal—would have been far too large for any meaningful optical follow-up observations. Still, LIGO's and Virgo's coordinates couldn't narrow down the event to a specific galaxy where the collision had taken place; they just indicated that it had occurred in a specific region in the sky—one about 150 times the area of the full Moon.

To shrink the observation window, Alexander, Berger, and their colleagues first opted for a powerful optical instrument, the Dark Energy Camera at the Victor M. Blanco Telescope in Chile. Operating the camera remotely from Harvard, they quickly scanned a very large area of the sky, one picture at a time. Within one hour, the team spotted a galaxy 130 million light-years away, with a new bright light source that hadn't been there before—galaxy NGC 4993 in the constellation Hydra. It was like an X very precisely marking the spot, said Berger. He later found out that the team determined the precise location within minutes of SWOPE, followed by the Distance Less Than 40 Mpc survey, the Las Cumbres Observatory in Panama, the Visible and Infrared Survey Telescope for Astronomy or VISTA in Paranal Observatory in the Atacama, MASTER in Russia, and a number of others.[47]

Excited, Berger phoned Metzger, the one who had theorized the existence of the kilonova in 2014. Metzger was beyond excited—especially when he found out that the kilonova's color and luminosity exactly matched his predictions, showing that the radioactive decay of heavy elements synthesized in the merger was under way. The color of the debris cloud was bright blue—meaning that it was extremely hot, like the ultrahot blue flames on a gas stove. The cloud lingered for several days, with telescopes all over the world monitoring its every shade. From blue, the color gradually shifted to a dull red as the ejected material cooled down. The astronomers could observe in detail the spectra (chemical fingerprints) of the kilonova, which showed that many heavy elements, including gold, platinum, and silver, had formed during the collision—solving the mystery of how these elements are created.

Berger's team wanted to see the breathtaking kilonova images in shorter wavelengths, too, so they submitted a proposal to the Hubble Space Telescope to take ultraviolet measurements. Every wavelength reveals something new, and at this point astronomers were trying to get as much information as they possibly could. Securing observation time with Hubble usually takes weeks, but these were extraordinary circumstances, and the researchers used "director's discretionary time" by submitting a brief proposal, just two paragraphs long (possibly the shortest proposal ever written, according to Berger). It simply said that they had found the first counterpart of a binary neutron star merger and needed to see it in ultraviolet light. It got approved—and they got the observations five days after the original detection of the gravitational waves. Nine days later, astronomers using the Chandra x-ray space telescope saw the first clear signals from the merger in x-rays.[48]

After the x-rays, at the very edge of the electromagnetic spectrum, are gamma rays—the most energetic radiation known. When the two neutron stars smashed into each other, most of the material combined and very likely collapsed quickly, under its own gravity, into a black hole. Some material, though, would have been blown outward at high speed, in the form of narrow jets of gamma-ray particles racing away at speeds close to the speed of light. This is what the Fermi Gamma Ray Space Telescope had detected less than two seconds after LIGO and Virgo had spotted the gravitational waves, as a very short burst of gamma rays.[49]

This type of gamma-ray flash was first spotted on July 2, 1967, by two US satellites, Vela 3 and 4. Researchers originally misidentified them as indications of possible nuclear weapon tests conducted by the Soviet Union. It took them a decade of observations to realize that these gamma-ray bursts originated in deep space and could last from a few milliseconds to hours; the ultra-short ones, less than two seconds long, were dubbed short gamma-ray bursts.[50] The burst now observed by the Fermi Telescope finally confirmed a long-standing theory that merging neutron stars might be the source of what researchers believe to be the most powerful explosions in the Universe—short gamma-ray bursts, or SGRBs.

Because the jet was seen at an angle, it looked different from the hundreds of SGRBs detected before—although the observations now showed just how different these flashes can appear when looked at sideways. Sifting later through archives of astronomy data, it dawned on researchers that they had observed very similar events in the past, but no one had been able to categorize them then.

Close to the other end of the electromagnetic spectrum are very long waves—radio waves. Neutron stars and black holes are traditionally observed with radio antennas, and dozens of these antennas were used to analyze the merger's aftermath. Kate Alexander was leading a team observing the radio spectrum with the Very Large Array in New Mexico—a cluster of telescopes made famous in the movie *Contact* with Jodie Foster. She knew from previous detections of GRBs that they start emitting radiation in radio and x-ray as the jet runs into the interstellar medium—the gas that permeates the space between stars and where the two neutron stars were spiraling before their merger.

At first, astronomers failed to detect any radio waves coming from the collision. Finally, a team led by Alessandra Corsi of Texas Tech University was the first to spot them, on September 5, 2017, a full sixteen days after starting their search. Alexander's team detected the signal too. The delay happened because the jet triggered by the collision was not pointing directly at Earth but traveling at an angle. Researchers continued to detect these waves for months, until they finally disappeared.[51]

Radio. Microwave. Infrared. Visible light. Ultraviolet. X-rays. Gamma rays. And gravitational waves. For the first time in history, multi-messenger astronomy was in full swing, observing two faraway, ultra-dense objects that had crashed into each other 130 million years ago, during their final moments in the Universe. The paper about the merger has more than four thousand co-authors—about a third of the world's astronomers. The data they gathered, from this merger and hopefully many future ones, will be analyzed for years to come.

There is still so much we don't know about neutron stars—and maybe, just maybe, observing these cataclysms from the vantage point of our blue dot in space will help us to shed some light on their inner structure, their peculiar jets that spit out particles and send radiation across the cosmos, their enormous magnetic fields that are stronger than anything else in the Universe, and many other secrets. We will never be able to travel to a neutron star, but with multi-messenger astronomy, which harnesses the power of various telescopes around the globe, and the help of giant particle accelerators, we will soon know more about them. Perhaps, too, we will be able to use what we learn about neutron stars to better understand the expansion of the Universe, find out more about the supermassive black holes lurking at the heart of galaxies, and finally determine once and for all whether Albert Einstein was really right with all his predictions of general relativity. Adding to the possibilities, high-energy astronomy, with its new x-ray telescopes like ROSAT, XMM, and Chandra as well as gamma-ray detectors like the Fermi Gamma-ray Space Telescope and now LIGO and Virgo, are allowing researchers to study these enigmatic objects in entirely new, exciting ways.[52]

As the aftermath of the distant cosmic neutron star drama was unfolding, Marica Branchesi's older son, Diego, was staring at his mother's computer screen in their living room. Suddenly he turned to her and—pronouncing every word very carefully and with much thought—said: "Mom, when you're done with the *binary neutron star merger,* can we go to breakfast?"[53]

DEEPER DIVE: The Origin of Gold

Where does our world come from? How do elements form? Every element on Earth was created one way or another in space. There are 118 elements in the periodic table, and 94 are found naturally on Earth. But right after the Big Bang, 13.7 billion years ago, none of them existed. There were only their elementary constituents, quarks—elementary particles that are usually found in triplets and make up the familiar neutrons and protons, which in turn make up atoms. The infant Universe was extremely hot and dense, and therefore couldn't bind the quarks; instead, they existed in a weird "soup" state for at least a few minutes. As the Universe expanded and cooled a bit, they assembled into protons (hydrogen nuclei) and neutrons, which then formed helium nuclei—made of two protons and two neutrons.

Expansion continued, and the temperature continued to drop—but it took 380,000 more years for electrons to get trapped in orbits around nuclei, forming the first very light atoms, mainly hydrogen and helium, and some lithium. Fast-forward another 1.6 million years, and gravity led to the formation of early stars and galaxies from clouds of gas, and to heavier atoms such as carbon, oxygen, and iron. Massive stars became giants, and in their cores they fused helium into carbon and produced magnesium, nitrogen, oxygen, neon, and iron, which marks the end of core fusion. When stars died in a supernova explosions, even heavier elements—like nickel, cobalt, copper, manganese, zinc, and vanadium—were formed.

But when researchers ran computer simulations, they realized that supernovas were not powerful enough to create elements heavier than iron. So where did all the silver, gold, platinum, mercury, molybdenum, uranium, and others come from? Scientists theorized that these elements might be created during the merger of neutron stars, in the so-called r-process. "R" here stands for rapid—the rapid neutron-capture process, a set of nuclear reactions during which neutrons are captured by heavy seed nuclei (like iron) to build up elements heavier than iron. During such a merger, neutron stars release a lot of neutrons, which, when heated to extreme temperatures, bombard surrounding atoms to form heavier elements. When the collision of two neutron stars was first documented, scientists could for the first time observe the radioactive decay of these heavy elements as they formed during the merger—as a blue kilonova.

How these heavy elements got to Earth is a different question. Some of these elements may have arrived in meteors. Nickel and cobalt, for example, are often found in iron meteorites: iron and nickel or cobalt are produced together during nucleosynthesis in supernova explosions. Alternatively, they may have been present in the matter that came together when the Solar System formed 4.5 billion years ago and were then released from the Earth's crust over time.[54]

DEEPER DIVE: Why Was the Kilonova Blue?

The color of an object depends on the wavelength of the light that it emits. Light behaves like a wave or a particle, under varying conditions, and the wavelength is the distance between the two successive wave crests (or two troughs). This wavelength varies depending on what part of the electromagnetic spectrum the light is emitted in: gamma radiation has the shortest wavelength, while radio waves are the longest. The energy carried by each individual photon—the basic unit that makes up light—is inversely proportional to the wavelength, meaning the shorter the wavelength, the more energy that type of radiation will carry. Gamma rays are extremely energetic, radio waves much less so, and visible light is somewhere in the middle.

Now if we zoom in at the visible light part of the spectrum, the wavelength is very short on its blue end and gets progressively longer as it goes to the other end—red. In other words, blue light carries more energy than red light. There are objects that absorb light perfectly, known as blackbodies. The Sun is a blackbody, and so is any solid or dense gas. It's possible to describe the spectrum of these objects mathematically, and the wavelength where more light is emitted is inversely proportional to the object's temperature—meaning the colder the object, the redder its color is, and the hotter it is, the bluer it is (the opposite of the convention of marking water faucets red for hot, and blue for cold). Some stars are blue—meaning really, really hot, about 7,000 degrees Celsius—and others are a cooler red, a mere 4,000 degrees Celsius. Go below 4,000 degrees, and the emission won't be detectable in visible light, although there will still be more emission in the red range than in the blue. On Earth, we associate heat with something that's red, like a campfire, but that's mainly because it's more difficult to get a flame so hot that it emits a blue light.

After the two neutron stars merged, optical telescopes registered a blue emission from the radioactive decay of heavy elements—the kilonova.

Researchers once thought that a kilonova resulting from a neutron star merger would be entirely red, because the heaviest r-process elements (those with an atomic mass number greater than 140 on the bottom of the periodic table) are very efficient at absorbing blue light and preventing it from escaping the ejecta of the merger.

But in 2014, Brian Metzger and fellow astronomer Rodrigo Fernandez argued that the kilonova would have separate blue and red components. Not all of the ejecta released from the merger and its aftermath, argued Metzger, would necessarily contain such heavy elements with an atomic mass number greater than 140. If only the lighter r-process elements, with an atomic mass number less than 140, were synthesized in a given portion of the ejecta, then the kilonova emission from that part would be blue. And, he said, because elements in the ejecta would be from different parts of the accretion, some with an atomic mass number greater than 140 (producing a red emission) and others with an atomic mass number less than 140 (producing a blue emission), both could be observed in the same event.

How blue the kilonova becomes depends on the composition of the ejecta and the number of neutrons and protons, which in turn depends on the lifetime of the neutron star after the merger, before it collapses into a black hole. The longer the neutron star survives, the greater a fraction of light r-process elements are created, and thus the bluer the kilonova. Its colors encode information about when the black hole forms.

Indeed, the kilonova observed during the merger was very blue—especially in its early hours. Some red emission showed up later—evidence, Metzger says, that the neutron star did not survive indefinitely. A black hole formed, he thinks, probably just a few hundred milliseconds after the merger.[55]

◆ 2 ◆

*Discovering Neutron Stars
...and Little Green Men?*

"THE PULSAR ARRAY is in a bit of a sorry state. All the copper wire has been stolen."[1] Malcolm Longair, a chirpy, white-haired, seventy-seven-year-old British astrophysicist, shakes his head in disapproval. Longair is not just any astrophysicist; he's the former Astronomer Royal for Scotland and head of the Cavendish Laboratory at Cambridge University, some twenty minutes' drive away. We are standing in a field in front of a dense, prickly, and annoyingly tall hedge. Behind the hedge once stood the Interplanetary Scintillation Array—a vineyard-like collection of a thousand cedar poles, each twelve feet high and connected by 120 miles of copper wire. It was this array that on August 6, 1967, picked up the first signal from a pulsar, confirming the existence of neutron stars. Until then, they had been just a theory.[2]

Longair shows me the site on a chilly January afternoon, after a brief visit to his place of work, the Cavendish Lab. That's where, in the late 1800s, physicist Ernest Rutherford started working on nuclear transformations, an effort that in 1911 would lead him to discover the nuclear structure of the atom. It was also where in 1932 British physicist James Chadwick discovered the neutron.[3]

Next to the hedge, on the other side of the defunct array, is a true graveyard of astronomy—four black radio telescopes on rails are motionless, their dishes like dried flowers stretching withered petals toward the sun. Next to them stands another equally derelict radio array, alongside an abandoned control room. Wooden poles sticking out of the ground are all that remains of the Interplanetary Scintillation Array—a piece of science history that became victim of a huge, temporary surge in commodities prices; just like many church roofs and other easy-to-access "sources" of copper, its wires were stolen and recycled with the help of unscrupulous scrap dealers.

Rewind half a century, though, and these fields were buzzing with eager graduate students and researchers. The 1960s was a fertile decade for radio astronomers. Caltech astronomer Maarten Schmidt solved the mystery of quasi-stellar radio sources, or quasars, in 1963, showing that they were distant hyperluminous active galaxies. Eventually, they were to be associated with supermassive black holes at the center of galaxies.[4] Two years later, US radio astronomers Arno Penzias and Robert Wilson at Bell Labs in Holmdel, New Jersey, accidentally discovered cosmic microwave background radiation, the leftover glow from the Big Bang.[5] And it was here in rural Cambridgeshire that a young graduate student from Northern Ireland, Jocelyn Bell, made history when in 1967 she spotted a weird signal that looked like a scribble—she called it a "scruff"—on miles-long rolls of green and white chart paper spooling out of a small pen recorder. The scruff turned out to be a faraway pulsar timidly saying hello. But she didn't know it at the time. No one did.

TO GET TO THE SCRUFF and understand its significance, we must rewind even more—another fifty years or so, to the cusp of the twentieth century. It was an amazing time to be a physicist: J. J. Thomson had just discovered the electron, in 1897, while at McGill University in Montreal;

Rutherford, who had moved there from Cambridge in 1898, was disentangling the sequences of radioactive decay chains. On his return to Manchester, he studied the scattering of energetic alpha particles by thin gold foil. In 1911, Rutherford presented his model of the nuclear atom. Atoms, he said, are mainly empty space, with a tiny nucleus consisting of positively charged particles surrounded by a swirling cloud of negatively charged electrons. But there was a problem. There was more mass in the nucleus than would be expected if the number of electrons were to balance the number of protons. So researchers assumed at the time that there were electrons in the nucleus that would neutralize the "excess" positive charge. In 1920, Rutherford proposed that, rather than nuclear electrons and protons, there might be a neutral particle—which he called the neutron. "For the next decade, Rutherford and Chadwick, who had transferred to Cambridge with Rutherford from Manchester, made strenuous, but unsuccessful, efforts to find the elusive neutron," says Longair.[6]

Finally, in 1932, Chadwick netted the neutron. He later scooped up the Nobel Prize in Physics for the discovery.[7] "We kept the instrument that Chadwick used," Longair says, pointing at an unassuming metal tube some 15 cm long at the Cavendish Lab, on display behind glass. Longair has long been taking care of a collection of old photographs and various instruments, used by researchers over the past century and a half or so, showing them to the public (mostly scientists or students) as part of a carefully maintained exhibit. Not everything is on display, though: in his office, Longair shows me Rutherford's heavy oak desk that he has decided to keep close.[8]

With Chadwick's neutron discovery, the atomic model had at last become more complete than it had ever been, and the world noticed. But among all the scientists keen to probe the atom and its innards even further, one researcher went on a total tangent. It was Bulgarian-born Swiss physicist Fritz Zwicky—a rather erratic, cocky character who in 1925 started working at the California Institute of Technology (Caltech) in Pasadena. He was easy to spot: one just had to look for a physicist suddenly dropping to the floor in a lecture theater to do one-armed push-ups or picking a fight out of nowhere—that was Zwicky. Little wonder that many of his colleagues dismissed him as an irritating buffoon. But it was Zwicky who jumped on Chadwick's neutron discovery and just two years later, in 1934, published (together with astronomer Walter Baade) a very short paper correctly suggesting a direct link between the death of a massive star, the ensuing supernova, and a neutron star left behind.[9] Later, Zwicky would go on to discover 120 supernovas.[10]

"It was just a wild speculation by Zwicky that you might be able to get a very compact star, because there was no electric charge to keep neutrons apart," says Longair. "That paper is no more than a tiny little thought, there's essentially no physics in it."[11] Still, Zwicky was right. Massive stars that run out of fuel to burn suffer the sad fate of core collapse under gravity, followed by a spectacularly explosive funeral—a supernova (see Chapter 3). What's left is a very compact yet incredibly dense object: a neutron star. Today, no one thinks of Zwicky as a buffoon; he is regarded as a genius not only because of his supernova-related insights, but also due to his work on dark matter and galaxy clusters.

Back in 1934 though, not many paid much attention to his insights. Astrophysicists at the time were much more concerned with understanding white dwarfs—the remains, we now know, of less massive stars that usually don't go supernova. Still, three years after Zwicky's paper another scientist, Soviet physicist Lev Landau, published his work on what he called the "neutron core" of a star.[12] Among the researchers who noticed Landau's paper was Robert Oppenheimer, who in 1939 went on to calculate the upper limit at which neutron stars can remain stable, estimating the probable mass of these objects before they have to collapse into a black hole. Oppenheimer and his student George Volkoff determined mathematically that this maximum mass should be around 70 percent of the mass of the Sun—an incorrect result, though, because they didn't take into account the strong nuclear force between neutrons.[13] Later theoretical work placed the limit at somewhere between one and a half and three solar masses. Oppenheimer "did all this work and made the first estimates of what the temperature should be of the surfaces, but these objects—neutron stars—were so tiny that everyone thought they were totally undetectable," muses Longair. X-ray telescopes didn't exist yet, so researchers thought that there was no way of ever finding them, which explains the lack of interest from the astronomical community. "What could you do about them at that time? Nothing. That's why although they were known as theoretical possibilities, they were more or less regarded as exotica," says Longair.[14]

AND THAT'S HOW things stood until after World War II. In the intervening years, Oppenheimer's attention got deflected to other—more, ahem, practical—work, such as developing an atomic bomb. Neutron stars were left to gather dust in the theoretical desk drawer until 1967, when they returned to scientists' attention with a bang.[15] "The 1960s were the key turning point in modern astrophysics," says Longair. "That's when all the new astronomy really started."[16]

Indeed, until about the 1940s, astronomers' observations of the Universe were mostly based on objects emitting light in or near the visible spectrum. The electromagnetic spectrum unites all the frequencies of electromagnetic radiation we can detect, ranging from the lowest to the highest frequency (which also means longest to the shortest wavelength)—from radio waves to infrared, visible light, ultraviolet, x-rays, and gamma rays. With radio astronomy, researchers suddenly gained an extra sense that they had never had before. It allowed them to discover whole new worlds, by tracing the radio waves they were detecting back to their source, then trying to look at that source object in visible light.

It all got a bit confusing, though, when in the 1950s telescopes began to pick up strong radio emissions from sources that did not correspond to any visible object in the sky. The sources were compact, yet "shone" brightly in radio. They were a mystery. Then in 1962 John Bolton, director of the Parkes Observatory in Australia, and astronomer Cyril Hazard applied the so-called lunar occultation technique to observe one of these particularly bright sources.[17] They used the known orbital path of the Moon to determine precisely where the source was when the Moon passed in front of it, blocking the radio emission. They then looked at the source again when the Moon moved past it, unblocking the radio wave and making it reappear.

The following year, Maarten Schmidt observed the source with the Hale Telescope on California's Mount Palomar and detected a visible jet of radiation streaming from it. Having analyzed the spectrum, the scientists noticed that it was greatly redshifted, meaning that the object was racing away from Earth at one sixth of the speed of light, about 31,000 miles per second (50 km per second). They also found that it was three billion light-years away—much farther from Earth, yet brighter, than many galaxies known at the time. Schmidt had detected the first quasi-stellar ("similar to a star") radio source, which today we know by its shorter nickname, quasar. The next year, in 1964, it was suggested that quasars are in fact supermassive black holes believed to be at the center of most large galaxies—that is, they are a type of what's known as active galactic nuclei.[18]

"It was a very dramatic period and some of us were just lucky to start doing research exactly when all of this was exploding. Very exciting times," says Longair.[19] He had joined the Cambridge radio astronomy group as a postgraduate student in 1963—and dove straight into searching for more quasars, at the request of seasoned radio astronomer Martin Ryle.

When Chadwick discovered the neutron, the Cavendish Lab was based in the center of Cambridge. Today, it's at the outskirts. But Longair's office in the old lab happened to be next to that of an astrophysics professor named Anthony Hewish, who was interested in quasars and especially their scintillation—the seemingly random fluctuations in the intensity of radio waves they emit.[20] It's the same effect as when we see the twinkling of stars—they only appear to twinkle because of turbulence in the atmosphere of the Earth, which the light from the stars travels through on its way to us. This turbulence makes the air lumpy, forcing the light to be focused or de-focused, and because the air is moving in the wind, the brightness of each star seems to rapidly fluctuate, or twinkle. As radio waves go through the interstellar medium, the very low density gas between stars, they experience similar interference. The interstellar medium is not uniform—it is denser in some places and more diffuse in others. Close to home, too, blobs or clouds of plasma blown out from the Sun—the solar wind—make a radio source appear to "twinkle." While quasars, as compact radio sources, do scintillate, radio galaxies (another type of active galactic nuclei) that have a larger angular diameter do not, and by studying how the scintillation of radio sources changes, astronomers can learn more about the lumpiness of the interstellar medium.[21] (For more on what is in the interstellar medium, see "Deeper Dive: The Interstellar Medium, Home of Neutron Stars.")

Hewish thought it should be possible to cherry-pick quasars from the sky by searching for scintillating sources using a large radio telescope. His colleague Martin Ryle had just finished a sky survey that had discovered several thousand radio galaxies, many of which may have been quasars. It was impossible to pinpoint which ones, though, because Ryle's telescope worked at too high a radio frequency to show the scintillation effect. Hewish had an idea: build an array that was highly sensitive but designed to capture very low frequencies instead.[22] He secured a grant from the UK Department of Scientific and Industrial Research for a mere £17,000.[23] It wasn't much, but it was enough to hammer a thousand poles into the ground in an open field in rural Cambridgeshire. Construction started in 1965.[24]

While Hewish was the brains behind the Interplanetary Scintillation Array, he wasn't about to swing any hammers to knock the required poles into the ground. That glorious job belonged to six graduate students. Among them was Jocelyn Bell, a rare woman on the physics faculty at Cambridge—there were typically one or two new female graduate students per year in radio astronomy, says Longair. The twenty-two-year-old had recently arrived from Northern Ireland and, having completed her undergraduate degree in physics at the University of Glasgow, started working at the Cavendish Laboratory in 1965 (although she was not on the faculty there, says Longair). Armed with a three-year grant to complete her PhD at Cambridge, like all other graduate students she mainly

Jocelyn Bell
(Roger W. Haworth / Wikimedia Commons, CC BY-SA 2.0)

worked in the attic of the lab—and in an open field. Building the array and making sure it worked as it was supposed to—studying the fluctuations of the radio source intensities, investigating the role of the solar wind, and searching for quasars—was the core objective of her PhD. It took the students two years to build the array, working year round, pushing through windy Cambridge winters. The telescope started taking data in July 1967.

Bell became very tanned and muscular from all the field work. "Colleagues at the lab would constantly ask me whether I had just returned from a skiing holiday," says Bell, who is now a soft-spoken visiting professor at Oxford University. As we talk, she doesn't elaborate on her answers very much, probably the result of decades of giving interviews. But her replies, as well as her email messages to arrange our meeting, are so lighthearted and twinkling with humor that she instantly puts me at ease.[25]

With its wooden poles, Antony Hewish's scintillation array was certainly different from the typical radio dish that most people think of when asked to describe a radio telescope. Still, plenty of early radio astronomy was done with simple arrays like that—a collection of poles and miles of copper wire. To make it work across the large collecting area, it was important to "phase it up," meaning one had to make sure that the signals arrived with the right delay from different parts of the array, so that they could all be joined up to make a single incoming signal.[26] It was Hewish who had pioneered the technique of measuring the scintillation of radio sources. When Bell started using the array, only about twenty quasars had been known; by the time her work was finished, she had discovered about another two hundred.[27]

The array was about 18,210 square meters (4.5 acres)—a tad smaller than two football fields next to each other. "That was enough to detect fluctuations in the intensities of the radio sources on the timescale of about one tenth of a second," says Longair.[28] Apart from miles of copper wire, Hewish also bought 13.5 km (8.5 miles) of cable and 124 km (77 miles) of reflector wire. The poles were positioned in sixteen rows and were set up to create the required phase delays between them.

The array was able to detect radio waves with a wavelength of nearly 3.7 meters, or twelve feet, meaning its observing frequency was very low, at 81.5 MHz. It worked like a typical television antenna—a spike of metal that picks up radio waves—but multiplied by 2,048, the number of dipoles. Unlike a steerable dish, the array scanned the entire sky as the sky moved over it. When a radio wave came in, its oscillation caused the electrons in the wire connecting the poles to vibrate. The vibration was then transmitted to a connecting cable linked to the receiver in the lab, where the oscillating electric current would give an output. Today, the cable would go into a computer as a string of numbers. Back in 1960s, it was analog—so Bell had to scrutinize miles and miles of squiggles drawn by a red pen on moving narrow rolls of chart paper that were continuously spooling out of her pen recorder.[29] Some of these rolls are now on display in the old control room near the field that once housed the array.

She was the only person sifting through the data because, she speculates, it was deemed not enough science to have more than one graduate student working on it. But it was extremely tedious and time-consuming, and, she adds, "I couldn't keep up with it. I had a backlog."[30]

To spot the scintillation and hence a quasar, Bell had to carefully look for changes in the intensity of the fluctuation. She had to make sure the squiggles on the chart paper represented radio waves from a faraway source and not from a nearby tractor or a passing car, which could produce interference that looked similar to radio waves. The squiggles from cosmic sources were colloquially called Chads—a specific interference pattern that has two small bumps and a big bump in between them. The name comes from a wartime cartoon with a character named Chad who would peep over a wall, with just his hands and head (with nose and two eyes) sticking out, asking about sugar or eggs or other staples that had just disappeared.

Hewish suggested to Bell that she make a map of all the sources in the sky, so every time she saw a scintillating source, she had to make a cross, by hand. If the source appeared in the same position each week, she would know it was a genuine source, a quasar. She produced this map every week, analyzing hundreds of feet of paper per day, with data coming off the telescope all the time.

On August 6, 1967, among four hundred feet of chart paper, she noticed a rather odd squiggle that puzzled her. It was less than a quarter inch on her readouts. A scruff, she called it, logged it with a question mark, and moved on to the next section of the paper. She knew, though, that something was way, way off—it looked like a strongly scintillating source in a region away from the Sun. But it didn't look like interplanetary scintillation.

Intriguingly, every time she observed it, the signal seemed to come from the same place in the nighttime sky. Scintillation, though, usually occurs during the day, because it's a solar-based phenomenon—and Bell had deliberately chosen to research this area because she hated staying up late at night. It was also weird that the "scruff" never took place in all three bits of the Chad. Sometimes there was a one-minute burst, or sometimes two, but never three.

Bell's PhD supervisor, Hewish, didn't supervise her very closely, rather expecting her to just get on with things. But he was there if there was something that troubled her. Once she realized this funny signal was coming from the same part of the sky again and again, she went to him to brainstorm.

Hewish was equally intrigued and decided to follow up. He told her to make high-speed recordings of the signal, running the chart paper faster—similar to making a picture larger. His idea was to get the signal to take up not just a quarter inch but to spread out more, to enable the researchers to study its structure. Bell started doing it every day for some time, but couldn't run the recorder nonstop, because the paper would quickly run out. Even more frustrating, the scruffy signal proved finicky—for a month, at exactly the times she was doing the recording, the source was too shy to come out to play. It seemed to have vanished. Bell was about to give up hope when it finally reappeared on November 26.[31] It was there; it was real. The next day, she saw that it was pulsating every 1.33 seconds—rather a small period for a star. Was it artificial?

For Hewish and Bell, the challenge now was to exclude every possible source of interference before announcing the discovery as a new cosmic radio source—because a mistake would surely affect their reputations. "A lot of the time went on trying to establish some basic facts about the thing, but also trying to find other things, human-made things that could explain that signal," Bell says. While the squiggle did look like some interference, it oddly kept to sidereal time—a timekeeping system used by astronomers to pinpoint objects in the sky. It's based on the Earth's rate of rotation relative to stars and makes a sidereal day slightly shorter than the usual day—23 hours 56 minutes. The new pulsating signal was being detected at the same time in each sidereal day. "I remember trying to work out if there could be a satellite in an orbit that would make it reappear every 23 hours 56 minutes, but I couldn't find a stable orbit for such a satellite," recalls Bell.[32]

After it became clear that the object was following this sidereal cycle, happening four minutes earlier each day, twenty-eight minutes a week, and going on for several months, Bell understood that it wasn't "Joe Bloggs driving down the road in a badly suppressed car," she says. "That's not the work pattern that humans keep to."[33] Hewish even wrote to all the observatories in Britain and asked them if they were running any programs since August that might have been causing interference.[34] They all said no. And because the Chad was not totally symmetrical, it was possible to say that the weird signal was going through the telescope at the same rate and in the same direction as the quasars she had been hunting for.

Hewish and Bell now asked a couple of colleagues, Paul Scott and Robin Collins, to check if they could detect the signal with their instruments. They didn't see it at first, but only because of an error in calculating when the source was supposed to transit their instrument. But then, finally, they did detect it—proving that the signal clearly wasn't due to some instrumental fault.[35]

For a few weeks, the researchers kept the odd squiggle to themselves—to the point that Longair, next door to Hewish and present at many common meetings, had no idea about the discovery. No one had, apart from a very tightly knit team of a handful of people. "There were rumors that something peculiar had been discovered, but what it was and what it was about, nobody knew," remembers Longair. The secrecy was partly due to the group being wary of theorists; they worried that theorists would steal their observations before they could analyze them, says Longair. Mostly, though, they kept it secret to make sure they had got it right and had done everything they could to exclude all possible sources of interference.[36]

To measure how far away the source was located, the researchers had to take into account a phenomenon called dispersion. When a radio signal propagates through interstellar space, it constantly bumps into free electrons zooming around. Once emitted, a signal is made up of different frequencies, and while they start off together, they are affected by these collisions differently. Higher frequencies zoom nearly straight through, but lower ones get delayed, arriving at our telescopes slightly later. Astronomers could estimate already back then the number of electrons in interstellar space, and measuring the time delay between the arrival of pulses told Bell and colleagues that the source of her scruff lived about 65 parsecs (or about two hundred light-years) away. That put it inside the Milky Way, but outside the Solar System, somewhere toward the constellation of Vulpecula.[37]

If it was clearly coming from space, could it be aliens? What if there was an alien civilization signaling to us from far, far away? The team thought about it long and hard; it was certainly an option they had to consider. Still, if there were aliens on some faraway planet circling a star, it should be possible to see changes in the pulses as their planet moved in its orbit. But they couldn't detect any changes. By now it was December, and the team was debating how to publish the paper, and whether to mention the possibility of aliens. Bell was less than thrilled that a bunch of aliens might be contacting Earth and hijacking her PhD project—after all, she only had half a year left until her defense.[38] Why would "little green men be using a daft technique signaling to what was, and probably still is, a rather inconspicuous planet?" she wondered, annoyed.[39]

Hewish and Ryle were also concerned about potentially causing massive panic if they were to announce to the world they had received communication from aliens. "If you find something like this, do you tell the Ministry of Defense first?" wonders Longair. "Or maybe it's a potential threat due to some nasty person, nasty country, after all this was during the Cold War . . . so does it have significance for national security? It was taken very seriously."[40]

Annoyed with the lack of consensus on the odd source's nature, on December 21—just before going home for Christmas—Bell went to the lab one more time. It was early evening and she wanted to look through her backlog of charts, which were still coming out at a pace of one hundred or so feet a day. At nearly 10 p.m., just before the lab was due to close, she gasped. There was another bit of scruff, similar to the first one, but in a different location—exactly opposite Cassiopeia A, a supernova remnant in the constellation Cassiopeia and a very bright radio source. Stray signals from Cas A prevented good observations. Intrigued, she went to the array to monitor when that part of the sky was due to transit. It was two in the morning when she got to the field, a frigid, starry night. The telescope's receiver system didn't work properly because of the cold. She breathed on it, she recalls, and swore at it—and managed to get it to work for five minutes, which was long enough for her to spot another string of pulses that lasted for 1.19 seconds. It looked very similar to the first pulsating source, albeit in a completely different part of the sky. What a huge relief—she knew right there and then that it wasn't aliens, because it was highly unlikely that two groups of them would be signaling us nearly simultaneously from opposite ends of the galaxy. "Whilst you might have one freakish thing, that one peculiar anomaly, finding two suggests this is a new category of thing," she says. "We still didn't know what it was, but it made it pretty clear that it was some new kind of astronomical object."[41]

Shortly thereafter, Bell got engaged—and then in early January found her third and fourth signals in the never-ending charts of data, after she came back to the lab from vacation. In February, the researchers wrote their paper and submitted it to the academic journal *Nature*.[42] Longair remembers Hewish giving a seminar a couple of days before the publication of the paper, getting on stage in a large auditorium in late February 1968. Everyone was excited and puzzled—and Hewish himself acknowledged that he didn't know what the signal was. The most plausible explanation he could give to the stunned crowd was that the source was a vibrating white dwarf.

The paper came out in March, with a tentative explanation as to what the mysterious pulsating signal from space could possibly be. When the media found out that the person who made the discovery was a student—and a woman—they were particularly excited. This was a time when there were even fewer women in physics than today—and Cambridge University had not even allowed women to become full members of the university before 1948. Bell had her photo taken standing, sitting, and standing pretending to examine scientific documents. One journalist even asked her to run waving her arms and looking happy, because, after all, she'd just made a discovery! Journalists zeroed in on what they thought were the most relevant questions, such as whether she was taller than Princess Margaret, and how many boyfriends she had at once.[43] But while Bell may have had more media attention than Hewish, it was her supervisor who got nominated for the Nobel Prize in Physics. Hewish received the award in 1974, together with Ryle.

"I think that the situation would have probably played out differently now," Bell says softly, and looks away. Over the years, she's been doing a lot of work trying to get women into science and to promote women in STEM fields (that is, science, technology, engineering, and math). "It's much more normal—still not fully normal, but much more normal to have women in science these days," she says. Bell got married shortly after submitting her thesis, and moved away with her husband, a local government official, who changed locations frequently. She stopped studying pulsars, but never really left the field—after switching to x-ray astronomy, she became part of the team that observed with the British-American Ariel 5 x-ray astronomy satellite. Despite having been bypassed by the Nobel Prize committee, she received numerous other awards, including the 2018 Special Breakthrough Prize in Fundamental Physics, and she was president of the Royal Astronomical Society from 2002 to 2004. Decades later, Bell is still frequently invited to give talks and attend conferences around the world.[44]

AS SOON AS Bell and Hewish's paper was published, on February 24, 1968, researchers the world over jumped at the problem to discover the sources of all these mysterious pulses. Neutron stars hadn't been Bell and Hewish's first candidates for the strange signal; they had suspected white dwarfs. But scientists knew that the period of one of the sources, 0.25 seconds, was too small for a rotating white dwarf. "That was the killer blow: you couldn't make white dwarfs rotate that fast," says Longair.[45]

It might just have worked for a vibrating white dwarf, though—so in their paper "Observation of a Rapidly Pulsating Radio Source," Hewish and his team tentatively argued that the pulses they had detected may have been due to oscillations of either a white dwarf or a neutron star.[46] Some bizarre object, surely—a pulsating star. The moniker "pulsar," though, came about later, a couple of weeks after publication. Anthony Michaelis, the science correspondent at the *Daily Telegraph*, asked Hewish what he proposed to call this new type of star and continued: "As it pulsates, would not 'pulsar' be appropriate?" Hewish replied: "Yes, that might be a good name for it." It stuck—and Michaelis wrote in his article on March 5, 1968, that the name pulsar (pulsating star) was likely to be given to the new object in the sky.[47]

But having a name did not solve the fundamental problem: what were these new pulsating stars? Neutron stars could have indeed been an option, but until then, no one really believed that they could ever be detected. Their theoretically predicted tiny sizes—the diameter of an average city—and the assumption that they wouldn't radiate heat, made them extremely tricky to spot, at least in optical or radio.

It's not that no one had been looking for pulsating radio sources. In 1951, Austrian astrophysicist Thomas (Tommy) Gold proposed the existence of such objects, in a paper titled "The Origin of Cosmic Radio Noise" that he had written for a Royal Astronomical Society conference in London; but no one listened. And just before Bell spotted that first scruff in her seemingly never-ending stream of charts, astrophysicist Franco Pacini, at Cornell University, submitted a paper to *Nature* that outlined a model of rotating and pulsating neutron stars. It was published in November 1967, while the Cambridge team was still guarding its discovery behind closed doors. Pacini theorized that a rapidly spinning neutron star with a magnetic field should emit a jet of radiation that we'd be able to observe. He also wrote that this radiation could end up in a supernova remnant around a neutron star—something Zwicky had first mentioned in 1933.[48]

A few months after the publication of Pacini's paper, Bell's discovery landed with a boom. Gold felt vindicated—and independently of Pacini, applied his earlier theoretical ideas about pulsating radio bodies to the new detection, arguing that the strange pulses were coming from a rotating neutron star. He thought that regular pulses were due to plasma in the neutron star's magnetosphere (a term he coined) being pushed to a fraction of the speed of light by both the high rotation speed and the strong magnetic field of the neutron star. Just as his reasoning had been dismissed in 1951, now in 1968 the world of science frowned at his theory once again. He wasn't even allowed to present his research at the first academic conference on pulsars in May 1968 in New York, to discourage, as the committee put it, ridiculous theories. Gold recalled the committee's response in his memoir: "Your suggestion is so outlandish that if we admit this for presentation, there would be no end to the number of other equally crazy suggestions that we would have to admit."[49]

That conference took place on May 20 and 21. Gold sent his paper to *Nature* the same week—and it was published on May 25. The editor at *Nature* clearly knew better than the organizers of the conference. Indeed, just a few months later astronomers discovered two young pulsars, Vela and Crab, residing inside supernova remnants—a clear link between a neutron star and a supernova (see Chapter 3). With Crab, astronomers also noticed the space between pulses ever so slightly increasing with time, which means that the pulsar is slowing down. It also means that it is rotating. Until then, it wasn't clear whether neutron stars rotated or vibrated. If something, for example a white dwarf star, vibrates, its vibrations get faster and faster as time passes. The new, slowing objects were clearly not white dwarfs but were most likely rotating neutron stars. Crab Pulsar is still perhaps the best-known pulsar clearly observable in its beautiful home—the Crab Nebula.

After Bell's discovery, astronomers the world over rushed to their telescopes to try and find more pulsars in the sky—and to make a dent in this new, exciting field of radio astronomy. One of the best pulsar discovery machines for decades was Parkes Observatory in New South Wales, Australia—which is also famous for being one of the handful of radio antennae used to receive live television images of the Apollo 11 Moon landing on July 20, 1969.[50]

Parkes, then, is where I decide to go next. With a bit of nostalgia, I leave behind rural Cambridgeshire and its derelict Interplanetary Scintillation Array, where the young Jocelyn Bell had boldly used a sledgehammer to build the future of radio astronomy.

GETTING TO PARKES is not trivial; like most major telescopes, it's in a remote location. I go by car from Melbourne—accompanying Matthew Bailes, his graduate student Renee Spiewak, a budding, home-schooled astronomer from Wisconsin who aims to be the next Jocelyn Bell, and a twelve-year-old boy prodigy, Rudra Sekhri, who loves turning pulsar emissions into music for fun, takes university physics summer courses, has written a four-hundred-page book on technology, and knows more about gravitational waves than most people ever will. He recently asked Bailes if he could occasionally come to Swinburne University of Technology and work with pulsar data, and is thrilled to go on this trip—despite the twelve-hour drive to get to the telescope.

Bailes is full of stories about pulsars, which makes the long drive rather entertaining—assuming you are fascinated by radio astronomy. Late on a Saturday afternoon, we finally arrive in the tiny town of Parkes, with a population of about 11,500. I am excited. This visit is like a belated birthday present for me; I had to celebrate my birthday all alone in the air, during my flight to Melbourne. Had we arrived at Parkes a decade earlier, we could have stayed on site, Bailes tells me, his voice filled with nostalgia. He much preferred those days, he says, remembering the thrill and adventure of being right next to the amazing instrument receiving signals from outer space. Nowadays, all observations at Parkes are done remotely; you can stay at home on your sofa and in your pajamas while you operate Parkes from your laptop.

A quick dinner in town, then a night in the small motel where the few astronomers who still come here usually stay; the next morning we set out on a twenty-minute drive to "the Dish"—the nickname Aussies have given Parkes, thanks to a 1999 movie that tells the story of how the observatory helped NASA to broadcast the lunar landing. "After the movie came out, the number of visitors skyrocketed," Bailes says. Australians love their Dish—and many visitors wonder whether the actors or real-life astronomers really do play cricket inside the antenna, just like it's shown in the movie. Short answer: no, they don't, and the ball used in the film was soft, so as not to damage the sensitive surface of the telescope.

I spot Parkes from a few miles away—a huge antenna glittering in the scorching sun of the Australian summer. Radio silence on site is mandatory, so phones have to be in airplane mode—a requirement that I will encounter again and again at all radio observatories I traveled to for this book. It's a crucial requirement, to minimize the amount of interference caused by terrestrial sources, which could mask genuine signals from space, prevent astronomers from spotting new objects, or even mislead them. As we drive through a small gate, prompting a flock of pink galahs to soar from a low tree branch high into the sky, Bailes suddenly takes a left turn, in the opposite direction from the Dish. "Observer's accommodation," reads a sign on a light pole. "That's where I used to spend weeks in the 1980s," says Bailes. Not anymore. Today, he comes here very rarely, he says, maybe to show a visitor like me around, or to take a graduate student to the telescope, to make radio astronomy more real. The last time observers stayed here was about a decade ago.[51]

Back in the 1980s, Bailes says, astronomers coming to Parkes had to book their accommodation far in advance. They ate on site, and Jeanette, the cook, was a rather assertive Australian who took the researchers' eating habits extremely seriously. "Everybody was terrified of her," says Bailes. "You didn't dare be late for any of the meals, or you'd face her wrath. If she cooked a meal, and you didn't eat it, she'd snarl at you in the morning." One day, Bailes and several fellow astronomers booked the accommodation, but forgot to say that they were not going to have dinner. They arrived late, to discover their dinners on a bench, cold and waiting to be reheated. "We knew that if we didn't eat the meals, Jeanette was going to kill us," recalls Bailes. The next morning, before the cooks arrived, one of the researchers who was due to go back to Sydney the same day took a plastic bag, scraped all the dinners into it, and dumped it off-site twenty kilometers away. "You wouldn't dare not eat Jeanette's meals," he says, with a chuckle.[52]

It's not just Bailes who is sad about those long-gone days. When we finally drive to the telescope, a medium-built man greets us with a broad smile. "John Sarkissian" says the label on his hard hat. His official title is operations scientist, and he's one of just a handful of staff who work here day in and day out. Later he will tell me how much he misses the olden days, the buzz of the astronomers in the control room, the friendly competitions of running up the dish, dashing up and down those narrow ladders at all times of the day and night—without helmets or other safety equipment in those days—to manually change a receiver at the very top of the dish in the focus cabin some fifty meters above ground. Back then, one engineer, Harry Fagg, used to brag he could do the ascent in three minutes, laughs Sarkissian. Now it's all controlled by computers.[53]

"Beware of kangaroo droppings," Sarkissian warns us. As it turns out, there are indeed hordes of kangaroos, of all sizes, big stocky males that you wouldn't want to face in a boxing match; smaller, gracious females, some carrying adorable joeys in their pouches; and awkward teenagers, with their long legs and a body not yet fully formed. There are snakes here, too, Sarkissian says—three of the world's deadliest ones, the eastern brown snake, the western brown snake, and the king brown snake—which makes me rather nervous. But then we don helmets to walk right under the dish and go into the control room.

Here in this round tower supporting the control room it almost smells like history, despite all the equipment being squeaky new. Sarkissian is busy helping out NASA, monitoring the Voyager 2 spacecraft that just weeks ago, on December 10, 2018, ventured into interstellar space, at last joining its twin, Voyager 1. The two probes have been braving the vastness of cosmos for the past four decades and are still going strong. "See this red peak?" Sarkissian tells me, pointing at his computer screen. "That's Voyager, saying hello."[54]

Despite not being part of NASA's Deep Space Network, Parkes and its parent organization headquartered in Sydney, CSIRO, often give the Americans much-needed assistance. After all, for a long while Parkes, with its sixty-four-meter dish, was the largest steerable telescope in the Southern Hemisphere. Later NASA built its own dish—one six meters larger than Parkes—near the Australian capital, Canberra. Still, Parkes was a crucial tool for a number of space missions, from the Apollo Moon landings to tracking rovers on Mars and observing the Huygens probe during its descent into the weird, oily, hydrocarbon world of Saturn's largest moon, Titan.

And of course, from the moment it started taking data in 1961, Parkes has been doing radio astronomy. As soon as Bell's LGM-1 pulsar (for "little green men") was discovered and confirmed, the Dish has been a magnificent pulsar detection and study machine. Together with the Lovell Telescope at Jodrell Bank, it has spotted more than half of all the pulsars known today. One of Parkes's achievements was the discovery in 1983 of the first extra-galactic pulsar, PSR B0529 – 66, in our satellite galaxy, the Large Magellanic Cloud (LMC). Four more were found a few years later, one in the Small Magellanic Cloud (SMC), two in the LMC, and another one nearby. And after Parkes got upgraded with a cutting-edge multibeam receiver, allowing it to search much wider areas of the sky, in 2001 astronomers found thirteen more—the only extra-galactic pulsars known today. (For more on the multibeam receiver, see "Deeper Dive: The Multibeam," in Chapter 4.) Another major achievement of the Parkes multibeam surveys was the detection of the Double Pulsar, PSR J0737 – 3039A/B—the only system that consists of two pulsars, orbiting their common center of mass, and set on an inevitable collision course. *Science* magazine called its detection one of the top ten scientific breakthroughs of 2004.[55]

As Bailes looks at the computer in the control room just underneath the dish, he sees that Dick Manchester, a CSIRO astronomer and veteran of pulsar research, is remotely observing a pulsar from Sydney.[56] "OK, we've got permission from Dick to observe," Bailes says, sounding slightly nervous. Half-jokingly, he adds: "You always gotta be a little bit scared of your supervisor. I'm in my fifties, and he hasn't been my supervisor since 1989—but when he tells me to do something, I go, 'Oh wow, I gotta do it.'"[57]

Later that day, Sarkissian invites me for a totally surreal ride in the giant Parkes dish. A colleague of his dips the dish all the way down, until its edge is nearly touching the ground. As I walk onto it, I can't help but think that I'm stepping into a giant metallic soup bowl. The scorching Australian February heat makes the comparison even more apt, except that thankfully the heat is all coming from above. The telescope operator starts driving Parkes up, slowly, and we go higher and higher above the tree line until we stop, the bowl now completely horizontal and resembling a futuristic flower reaching for the sun. I walk up to one of the ladders that astronomers climbed up for decades; being here is incomparable to any other thrill I have ever experienced.

The night begins to descend, the sun slowly turning red, and then the birds come out. Tens, hundreds of them, large white sulfur-crested cockatoos, bright pink galahs, cheeky apostle birds, currawongs, and white-winged chuff, and the more common Australian magpie. At this time of day, the telescope is theirs. Packs of kangaroos hop all around us, with their tails and powerful legs looking like prehistoric raptors straight out of Jurassic Park. A few more moments, and the sun sinks below the horizon; we step off the dish. It's time to leave this place. Goodbye, Parkes.

DEEPER DIVE: The Interstellar Medium, Home of Neutron Stars

"Gee, this is really happening—I'm measuring pulses that have taken thousands of years to get to us," thought Jim Cordes, looking at the oscilloscope in front of him.[58] Blip-blip-blip, the signal went, pulsing up and down like a zig-zag line moving across a heart rate monitor. What he saw were pulses from a faraway cosmic lighthouse, a rapidly spinning neutron star immersed in the tenuous atmosphere of ionized gas and dust called the interstellar medium, which permeates the space between all stars (and neutron stars). While we do know quite a bit about this environment, the interstellar medium is still full of mysteries.

The year was 1972 and it was the first time that Cordes had traveled to the middle of the rainforests of Puerto Rico. Glancing away from the oscilloscope, he looked through the big window in front of him; just outside, a few dozen yards away, was a huge dish in the ground—the Arecibo Telescope. "You can see these effects from the interstellar medium and it gives you a sort of a visceral feel . . . and it just kind of connected me to it, just knowing that I was at the receiving end of all this," he says, sitting in his office at Cornell University. A pulsar astronomer, he became so intrigued by the effects of the interstellar medium on the signals emitted by pulsars that he has been studying it ever since.

Pulsars are ultra-accurate clocks on the decadal timescale—so reliable that they are even being considered as possible beacons to support navigation in space. Take millisecond pulsars, which spin at ultra-rapid speeds, rotating hundreds of times every second. Their pulses are extremely precise, because their extreme rotational velocities and high masses mean that they are very difficult to slow down; as a result, they send out a consistent pulse for a nearly indefinite time, braking by mere milliseconds after even billions of years. Because they are so reliable, even the slightest variation in a pulsar's data can be an indication of how the environment around this neutron star—the interstellar medium—is changing.

Contrary to what many people might think, the interstellar medium is not a perfect vacuum. It's made up of charged electrons and protons swirling around, about one for each cubic centimeter. The medium is also magnetized; its all-pervading magnetic fields can be found throughout the cosmos, at varying strengths. On average, the value of the magnetic field of the interstellar medium is a few microgauss—one millionth the strength of the magnetic field measured at the surface of the Earth, which is less than one gauss. In an MRI machine, the magnetic field is about 10,000 gauss. At the other extreme, the magnetic field of an average neutron star is around 10^{12} (one trillion) gauss. That's so strong that atoms get stretched into cylinders along the magnetic field lines.

When a pulsar emits radiation, it does so in a very wide range of frequencies across the electromagnetic spectrum. High frequency waves have a very short wavelength, so they aren't easily disturbed by the particles in the plasma; they go straight through the interstellar medium. But the lower frequencies keep bumping into electrons, which makes them lag behind; as a result, they arrive at the telescope later—a phenomenon called dispersion. The difference in arrival time for different frequencies depends on the number of electrons between us and the pulsar; it's a delay that may run to about one second. For a pulsar a thousand light-years away, which means that its pulses should take a thousand years to arrive, a one-second delay may seem negligible. But when scientists observe a pulsar, it's important to correct for this effect, to ensure that they can analyze all of the frequencies of the emitted radiation in sync.

Measuring the dispersion effect tells astronomers how many electrons are floating in the interstellar medium between us and the pulsar, which in turn gives them a handle on its distance to us. A more distant pulsar will have more electrons to ram through, meaning that the dispersion effect—or spread in arrival times—will also be bigger.[59]

This is not just about counting electrons floating in space. By understanding the properties of the interstellar medium, scientists can shed light on the formation and evolution of stars and galaxies. If they observe the same pulsar a year later and notice that the amount of dispersion has changed, it means the electron content in that region has changed, because there are areas of turbulence. The scintillation (or twinkling) of radio waves makes it possible for astronomers to study how clumps of material are moving between us and the pulsar. "We can measure that twinkling," says Cordes, "and work backwards and say, 'Okay, what kind of turbulence needs to be in the interstellar medium to produce that effect?'"

The interstellar medium briefly made headlines in August 2012, when NASA's vintage Voyager 1 probe, launched in 1977, left the Solar System—and ventured into this cold, quiet space between stars. On November 5, 2018, it was joined by its twin, Voyager 2—the craft that John Sarkissian still tracks from time to time using Parkes. Both probes carry a gold-plated record with video and audio about our world, just in case aliens intercept them. While Voyager 1 couldn't measure the properties of the interstellar medium because its plasma science instrument was damaged, the second spacecraft saved the day five years later—and found that the interstellar plasma was higher in density but colder and slower moving than the plasma inside the heliosphere, the region of space surrounding the Sun.[60]

◆ 3 ◆

When Stars Go Boom

I'M STANDING IN an endless carpet of flowers, mesmerizing in purple and pink, that extends all the way to snow-capped mountains on the horizon. This is supposed to be a desert, but after a brief, rare rainstorm flowers are blooming everywhere. It's hard to believe that this is one of the driest places on Earth.

Chile's Atacama Desert stretches thousands of kilometers across an arid, desolate, and high-altitude plateau, bordered to the west by the Pacific coast and to the east by the Andes. The oldest desert on Earth, it covers 105,000 square kilometers (more than 65,000 square miles) that are sprinkled with just a few mining towns full of workers extracting copper from this red-orange, rocky landscape. Now and then we drive past long-abandoned ghost villages, houses with empty windows gazing onto the vastness of this land.

The best-known tourist attraction in this desert is perhaps San Pedro de Atacama, an oasis with mud huts that provides a base for excursions to the nearby salt flats and salt lakes. As we drive from the Calama airport up the Pan-American Highway, I notice that this desert has hardly any sand. Yes, there are a few sand dunes in the Atacama, but not along the highway. The dunes, I hear, are popular with the tourists coming to San Pedro. Official tours take them by the van-load to the bottom of the dunes. After climbing to the top in scorching heat, the tourists, with plenty of sand in their shoes and socks, surf down on sandboards for the thrill of a lifetime.[1]

Outside the villages, not much thrives in this remote place. Some areas haven't felt a raindrop in years. Until 2017, the heart of the desert had been rain-deprived for half a century; when the downpour finally came, a team of astrobiologists discovered that the rain was causing a mass extinction of indigenous microbial species.[2] We pass huge rocky structures, as tall as mountains, that look as if they've just been through a snowstorm, but are actually covered in salt. The Atacama is an old landscape. Most rocks on the Earth's surface are, in geological terms, quite young, maybe hundreds of thousands or a few million years old. In the Atacama, some rocks date back fifteen million years. But it's not geology that has attracted me here. It's astronomy.

Several large observatories dot the vastness of the Atacama, one of the best places on Earth to look at the stars thanks to its remote location and unusual geography. It's so far away from any major towns that light or radio wave pollution is barely a problem. Clouds are also rare visitors to the Atacama skies; and given the altitude, this place offers as clear a view of celestial bodies as possible, short of actually going into space. All this is great for optical telescopes such as the Very Large Telescope (VLT) at the Paranal Observatory that I visit first—a futuristic-looking place that made a brief cameo in the James Bond film *Quantum of Solace,* with the housing accommodations for the telescope's scientists and engineers briefly transformed into a fictional eco hotel in Bolivia.[3]

For the telescope that I'm visiting this time, however, the dryness and altitude of the Atacama are the key attractions. I've come to see the world's largest observatory for the detection of ultra-short millimeter and submillimeter radiation, constructed here far away from the humidity or water vapor in the air that otherwise would absorb these wavelengths like a sponge. Its name is ALMA, short for the Atacama Large Millimeter/submillimeter Array, a network of sixty-six snowy-white radio antennas set on the Chajnantor plateau, high up on the northern end of the desert, at an altitude of 5,000 meters (16,400 ft).[4]

I've seen the array in pictures. It's an alien-looking, but somehow adorable cluster of small dishes that resemble mushrooms with weird, inverted caps. Before we get there, however, I must pass through ALMA's base camp, called the Operations Support Facility, some 2,000 meters (6,560 ft) lower than the plateau. First, I'm shown to a tiny cabin, a room six feet square with just enough space for a single bed and a shower; next up is a quick stop in the canteen followed by a safety briefing. That's especially important, because the next day we are to travel to the array itself. There, at an altitude of 5,000 meters (16,400 ft), oxygen is scarce and altitude sickness is no joke. Aldo, one of the technicians, warns that people have fallen ill before, suffering from nausea and debilitating headaches. If you are not careful and don't recognize the signs, you may die, he warns, as he shows me how to use the oxygen mask that I will be issued later. Next, I am given a medical exam, checking vitals like blood pressure and blood oxygen levels. Some visitors don't pass the test and aren't allowed to continue to the Chajnantor, but I guess my years of going to the gym finally pay off—I pass.

Early the next morning, with supporting staff, I head up to the plateau, clutching an oxygen mask. In my pocket I feel the coca leaves that locals swear by as a remedy for altitude sickness. I know we won't be staying for more than a couple of hours at this altitude, but I'm nervous. On the way up, we pass giant cacti, some reaching heights of seven meters (20 ft) or more. A llama and a couple of donkeys wander by, idly looking at our sluggish car. Finally, far in the distance, I spot our destination: the sixty-six huge dishes of ALMA, all working in unison to unlock some of the most stubborn secrets of the Universe. Tiny-looking humans in oxygen masks scurry around the bases of the dishes, operating and maintaining the array of antennas. Up close, the dishes don't look as adorable anymore, but so overwhelming that I gasp for air. Maybe it's just the lack of oxygen kicking in.

ALMA is not an obvious place for observing neutron stars, because the millimeter and submillimeter wavelengths are much shorter than the radio waves generated by pulsars. The array is normally used to study star formation, but it was ALMA that helped astrophysicists witness for the first time what they think was the birth of a neutron star. Scientists have a somewhat silly nickname for the supernova with the presumed neutron star inside: they call it "the Cow." Officially, it's called AT2018cow, the result of a naming convention that combines the detection year with a particular, preset sequence of letters—but the moniker stuck.

The lead author of the study investigating the Cow is Anna Ho, a graduate physics student at Caltech. When she recounts the day she first found out about the Cow, her voice lifts with enthusiasm.

On June 17, 2018, Ho received a mass email, along with hundreds of astronomers. It was from Stephen Smartt, an astronomer at Queen's University in Belfast. He and his team had just taken a closer look at a somewhat peculiar transient—a name given to an event that involves short-lived bursts of energy, say when a cosmic body suddenly changes state, like during a supernova. The event was detected a day earlier by the Asteroid Terrestrial Impact Last Alert System (ATLAS) in Hawaii—an automated astronomical survey and early warning system designed to spot small asteroids just before they hit Earth. The transient event was bright—very, very bright actually, between ten and a hundred times brighter than the typical blast of an exploding star. Ho, however, noticed something else about the object that made her jump up and sprint to her adviser's office: this supernova reached its peak brightness unusually quickly, within hours, while typical supernovas can take weeks to unfold. Around the world, telescopes, as well as several other instruments, were set to swivel and zoom in on the unusual event; among them the mighty twin Keck telescopes on Hawaii, and the Liverpool telescope on La Palma (one of Spain's Canary Islands).

At first, Ho's adviser and other colleagues looked for more mundane explanations for the unusual flash. Maybe it wasn't an explosion, they said, but just a bright star very close to Earth, in our own galaxy, fooling them into thinking that it was an explosion. Her adviser told her that it was definitely a star, and she shouldn't bother working on it at all.

At that precise moment, Ho's phone pinged; as she was walking to leave the room, she stopped to check her email and read the first confirmation that the flash was extragalactic, from around 200 million light-years away, in a dwarf galaxy in the constellation Hercules. "That meant that it was truly, truly an explosion! And that's when everyone got very excited. I just turned around, held up my phone to my adviser's face, and we all started running around frantically trying to figure out what to do about this thing," Ho says.[5]

With this explosion, for the first time ever, scientists might be able to witness a stellar death in real time—well, excluding the 200-million-year delay that it took for the light of the event to reach our planet. If the astronomers got their observations right, they might not only see a massive star going boom, but also watch its core collapsing, with a neutron star at its heart being born right there and then. That's why Ho sidestepped the typical radio or optical telescope option and chose ALMA to observe this particular supernova.

But to understand how Ho's quest really got started, you need to go back eighty-eight years, to a steamship sailing from Bombay Harbor, India, on its way to England.

THE JOURNEY ON BOARD THE STEAMER *Pilsna* lasted eighteen days. The year was 1930, and Subrahmanyan Chandrasekhar was traveling to Cambridge to start his postgraduate studies in physics. To pass the time, he toyed with equations. An Indian child prodigy (and nephew of Sir C. V. Raman, the first Asian to win the Nobel Prize in Physics, also in 1930), Chandrasekhar had graduated with a degree in physics by the age of nineteen. For the work he did on this ship, Chandrasekhar would himself receive the Nobel Prize, albeit half a century later, in 1983.

Shortly before his trip to England, Chandrasekhar had become fascinated by white dwarfs, the very faint corpses of stars. Scientists thought back then that all stars, including our very own Sun, would one day become white dwarfs, once they had run out of the hydrogen "fuel" that powered them. We now know that not all do—but those that do die this way gradually shed all their outer layers, until all that's left is a dense core of carbon, oxygen, and nitrogen. Eventually, it is thought that after a hundred million billion years, such a white dwarf will cool completely and stop emitting any light and heat, becoming a more quiescent black dwarf.

When Chandrasekhar boarded the steamer, astronomers had detected only three white dwarfs. Among them was Sirius B—the faint, dead companion to the bright star Sirius. It was already known that these white dwarfs had a remarkably high density—more than a million times the density of the Sun. Quantum mechanics, in its infancy in the early twentieth century, helped to explain how it was possible to reach such astonishing density. The gravitational pressure inside a dying star compresses the atoms in its core, squeezing them so tightly that their electrons get ripped away. That leaves the newly formed white dwarf composed of positively charged ions in a sea of electrons. Under gravity's continuing squeeze, quantum mechanics kicks in. One of its laws, the Pauli exclusion principle, states that no two fermions (for example, two protons or two electrons) can occupy the same state—just like when you play musical chairs, no two people are allowed to sit on the same chair at the same time. That means that inside a white dwarf some electrons are forced out of their low "ground" energy states into higher ones; in the process they generate what's called electron degeneracy pressure. It is this pressure that counterbalances gravity and keeps a white dwarf from collapsing in on itself.

Chandrasekhar knew all this, not least because he had closely studied *The Internal Constitution of the Stars,* a book written by Arthur Stanley Eddington, one of the world's most prominent astrophysicists of the time. Published in 1926, Eddington's book had brought white dwarfs to popular attention, even though the term had been coined four years earlier, by the Dutch American astronomer Willem Luyten. Eddington for his part wrongly explained the huge interior density of these dying stars as having to do with thermal (heat-induced) pressure on the atoms inside white dwarfs. The correct—quantum mechanical—explanation was put forward later that same year by English physicist Ralph H. Fowler in his article "On Dense Matter," published in the *Monthly Notices of the Royal Astronomical Society*. Chandrasekhar had meticulously studied this article as well, and when he boarded the steamer bound for England to study under Fowler, he decided to take the calculations further.[6]

The long voyage gave the Indian teenager the tranquility he needed to review Fowler's equations and expand on them by taking account of relativistic effects that come into play when objects move at speeds close to the speed of light. Chandrasekhar had correctly realized that the electrons inside the white dwarf would be zooming around at such high speeds, and the result of his equations was both unexpected and astounding. He found that for a star to become a white dwarf, the mass of its core at the moment of the star's whimpering death had to be less than that of 1.4 Suns. A white dwarf simply wouldn't be able to exist beyond that mass limit, he calculated, because any more mass would mean that the ultra-dense matter inside would not be able to resist the enormous squeeze of gravity. In other words, any star that by the end of its life was more massive than 1.4 Suns could not become one of these newly discovered, faint, yet super-dense objects in the sky. This calculation would later become known as the Chandrasekhar limit. But it also raised a question: what happened to all the stars that were more massive?

At the time, researchers didn't know. In 1931, when Chandrasekhar published his work, the neutron hadn't been discovered yet—Chadwick would detect it the following year. That's why the young Indian researcher had no idea what would become of more massive stars at the time of their death. Not knowing of the possibility of neutron stars, he suggested that they might—without gravitational pressure—simply shrink into near-nothingness. At the time, black holes, where very massive stars are thought to collapse, had been theorized but for decades had remained mere mathematical curiosities.

Unknowingly, Chandrasekhar became the prophet of both.

Eddington, for his part, never accepted Chandrasekhar's idea. After the initial publication of his equations, the young scientist submitted in 1934 two more papers to the Royal Astronomical Society that improved on his earlier calculations and ideas. The institution then invited him to give a talk about his results in January 1935. Eddington listened to Chandrasekhar and then took to the stage himself—to publicly ridicule Chandrasekhar's work. "The formula is based on a combination of relativity mechanics and non-relativity quantum theory, and I do not regard the offspring of such a union as born in lawful wedlock," Eddington said, to the gasps of a shocked Chandrasekhar. Eddington believed that any star would eventually become a white dwarf and found the idea of a gravity-driven collapse into nothing (what would later be known as a black hole) preposterous. "I think there should be a law of *Nature* to prevent a star from behaving in this absurd way!" Eddington exclaimed. Chandrasekhar was so shaken that he stopped working on white dwarfs for four decades. Most scientists sided with Eddington, given his preeminent status at the time. Still, despite the controversy, the two researchers managed to stay on good terms.[7]

THREE YEARS BEFORE Chandrasekhar was appealing to physicists to set aside Eddington's criticism and try to understand his ideas and calculations, a different scientific battle was breaking out on the other side of the Atlantic. It was early 1932, and Fritz Zwicky at Caltech had just heard about James Chadwick's discovery of the neutron. The news was a bombshell; most major newspapers around the world reported Chadwick's findings—after all, they would completely change Rutherford's model of the atom. "Discovers Neutron, Embryonic Matter," ran the headline on the front page of the *New York Times* on February 28, 1932. The accompanying article opened with this fanfare: "Dr. James Chadwick, working in Cavendish Laboratory at Cambridge, has discovered the neutron—one of the ultimate particles of nature. The discovery, first made public today, was

Fritz Zwicky
(Courtesy of Fritz Zwicky Foundation)

hailed by scientists here as the most important achievement in experimental physics since Lord Rutherford demonstrated the nuclei structure of the atom in 1911."[8] For Zwicky, the timing was perfect. Astronomers were already familiar with so-called novas—stars that would suddenly shine with increased brightness. They had also observed a much rarer type of nova, which was way brighter and surrounded by weird-looking nebulas. Zwicky called them "supernovas."

Zwicky and his fellow researcher Walter Baade had already speculated that supernovas might be the result of powerful star explosions. Now with the neutron in Zwicky's astrophysical arsenal, he connected the dots. What if, he argued in 1933, a star's core at the end of its lifetime collapsed under gravity's squeeze, and its innards became pure neutrons (as the protons captured the electrons and morphed into those neutrons)? This implosion of a "neutron star" would greatly reduce the original core's size and mass. Zwicky applied Einstein's famous mass-energy equivalence relation ($E = mc^2$) and suggested that the mass that was "lost" during the collapse of the core would convert into the energy that powered the explosion—in a supernova. In one of their two papers on the subject,

Zwicky and Baade proposed: "With all reserve we advance the view that a super-nova represents the transition of an ordinary star into a *neutron star,* consisting mainly of neutrons. Such a star may possess a very small radius and an extremely high density."[9] The two physicists presented their results at a meeting of the American Physical Society at Stanford in December 1933, two years before Chandrasekhar presented his research on white dwarfs.

Today we know that Zwicky's conclusions were correct, but at the time they appeared to be intuitive at best, since his papers contained few calculations to back up the claims. When Zwicky and Baade published their paper on supernovas and neutron stars in January 1934, the reception was rather cool, to put it mildly. For years, fellow scientists dismissed the concept of neutron stars as speculation.

On the upside, Zwicky's explanation for supernovas as a catastrophic event at the end of the life of massive stars was quickly accepted. It seemed to make sense, compared to the quiet whimpering deaths of lower mass stars, which were known to turn into white dwarfs. (For more on the demise of these behemoths, see "Deeper Dive: The Death of Massive Stars.") After all, supernovas had been on scientists' minds for a very long time—observed and recorded for many centuries. In 185 CE, early Chinese astronomers reported in wonder the appearance of a new star in the sky that outshone all its neighbors, but then vanished eight months later. The Chinese astronomers carefully recorded their observations and called the visitor—somewhat romantically—a "guest star." Today we think the event was probably supernova SN 185, in the direction of Alpha Centauri, somewhere between the constellations Circinus and Centaurus. In the following centuries, several more of these heavenly guests made an appearance, duly observed by Arab, Chinese, Egyptian, Japanese, Italian, and Swiss astronomers. These odd "stars" were observed twinkling in one place like all the other more permanent stars, and very much unlike another heavenly apparition zooming across the heavens, the occasional comet. In 1054 CE, a "guest star" shone for twenty-three days in daylight and 653 days at night and was so intensely bright that at its luminosity peak it is thought to have been four times brighter than Venus, which is easily identifiable with the naked eye. Later, it was recorded as supernova SN 1054 in the constellation of Taurus. This supernova is famous for producing the Crab Nebula—spectacular, multicolor gas that looks like a huge cosmic cloud.

Early astronomers, however, hadn't the slightest idea why these "guest stars" would come and go, until Danish astronomer Tycho Brahe spotted a new star as bright as Jupiter, on November 11, 1572, in the constellation Cassiopeia. He knew for certain that this star hadn't been there before. If nothing else, to Brahe it was proof that the sky was not immutable. A few decades later, in 1604, Johannes Kepler recorded another new star and came to the same conclusion. It was becoming clear that the heavens were not a crystal sphere or a piece of black velvet with shining dots attached to it. Still, it took another three centuries until Zwicky and Baade identified these lights as emanating from gigantic explosions signaling the death of massive stars.[10]

The neutron star part of Zwicky's theory, though, was not readily accepted until 1937, when Lev Landau, a Soviet physicist, wrote a paper proposing that all stars, including our Sun, had a neutron core "where all the nuclei and electrons have combined to form neutrons" that counteract gravity's squeeze, preventing their collapse. At first, Landau's work received little attention, because he had written it in Russian and published only in the *Proceedings of the USSR Academy of Sciences.* At the time, Landau was in real danger of being arrested, caught up in one of the purges instigated by Stalin during the Great Terror of the 1930s when millions were murdered. In an attempt to protect Landau by popularizing his work, the famous Soviet physicist P. L. Kapitsa had Landau's paper sent to Niels Bohr, a scientist in Copenhagen who in 1922 had won the Nobel Prize for Physics for his work on the atomic structure and quantum theory. Bohr was so impressed that he submitted the paper to *Nature,* which published it on February 19, 1938. The public recognition wasn't enough to prevent arrest, though, and the following year Landau was accused of "anti-Soviet activity" and thrown into the Butyrskaya, a notorious jail for political prisoners. He spent one year in the Butyrskaya and might well have died there, had he not been released thanks to Kapitsa's many letters petitioning Stalin.

Landau's paper came out a few months before the German-American nuclear physicist Hans Bethe would correctly and for the first time describe the mechanism powering stars: nuclear fusion. His work would later win him the Nobel Prize. Ironically, in its essence Landau's idea of neutron cores was wrong, but it revived the discussion about the existence of neutron stars. One of the readers of Landau's paper was US physicist Robert Oppenheimer, now prepared to consider the existence of these leftover cores of massive stars. Together with his graduate student George Volkoff, a Russian émigré from Vancouver, he began to calculate the theoretical maximum mass that such a neutron star could have before succumbing to gravity's squeeze. In effect, Oppenheimer tried to do for neutron stars what Chandrasekhar had done for white dwarfs: find an upper mass limit.[11]

During 1938 and 1939, Oppenheimer, Volkoff, and American physicist Richard Chace Tolman published three landmark papers that explain how neutron stars can form, what maximum mass they can have, and what happens if they go beyond this mass limit. They determined an upper mass limit of about 0.75 solar masses and explained that a neutron star, appearing out of the core collapse of a massive star at the end of its life, remains intact as long as its weight is counterbalanced by the short-range repulsive interactions between neutrons. But once a neutron star's mass gets too big—for example, when the debris from its supernova explosion falls back to the neutron core—the neutron star should collapse further and turn into a black hole. Oppenheimer and his colleagues, though, thought that neutron stars (and black holes) would be impossible to detect because they are so tiny (x-ray astronomy didn't yet exist).[12]

The upper mass limit of neutron stars is still known as the Tolman-Oppenheimer-Volkoff limit, although the limit they calculated was much too low, because they didn't take into account the strong nuclear force that keeps protons and neutrons in atomic nuclei. In the 1990s, with better technology and observations, the limit was raised to between 1.5 and 3 solar masses, until the LIGO-Virgo observations of the first neutron star merger placed the upper limit at approximately 2.17 solar masses. The most massive neutron star discovered to date weighs 2.14 solar masses—it's called J0740+6620 and is an ultra-rapidly spinning pulsar in a binary system, which means that it is orbiting a common center of mass with its companion star, a white dwarf.[13]

Oppenheimer and his student did their work in the tumultuous years right before World War II. As hostilities broke out, many physicists were forced to turn their attention to more pressing issues. Oppenheimer, for his part, dedicated himself to the Manhattan Project, working on developing an atomic bomb faster than Nazi Germany. For more than twenty years, neutron stars were more or less forgotten—until 1967, when Jocelyn Bell discovered the first pulsars. After this breakthrough, physicists quickly realized that it was possible to find neutron stars after all.

Bell's serendipitous discovery set the astronomical community abuzz. But the man who first proposed the idea of neutron stars and their connection to supernovas and the death of massive stars, Fritz Zwicky, remained strangely detached. Ron Ekers, currently an astronomer at CSIRO in Australia, remembers working with Zwicky at Caltech at the time.

I meet Ekers at the CSIRO headquarters, about twenty minutes from the center of Sydney. "Zwicky had an office in the basement of the same building as me," remembers Ekers. "I liked Zwicky and we often interacted, but he did not participate in any discussions of the neutron star-pulsar connection that I can remember. Many of us were talking about it at Caltech after the pulsar discovery, but for some reason he never got involved."[14]

By then Zwicky had already moved on to other things. He was one of those rare physicists who tackle a broad range of fields in physics throughout their careers. Besides being famous for his insights on neutron stars, he is also hailed for noticing that the bulk of the matter in the Universe appears to be missing. (For more on Zwicky's discovery of dark matter, see Chapter 6.)

In 1968, though, just as Bell and Anthony Hewish published their LGM-1 paper and scientists swiftly began suggesting that rotating neutron stars could be the source of these intriguing pulses, Zwicky was mostly preoccupied with finishing his work on describing the galaxies that had been observed. His *Catalogue of Galaxies and of Clusters of Galaxies*, running to six volumes and published by Caltech, listed 29,418 galaxies and 9,134 galaxy clusters that he had painstakingly identified.[15]

Another possible reason for Zwicky's lack of interest in neutron stars at this point, says Ekers, is that they suddenly were no longer theory. Researchers were now modeling these newly discovered objects. And as Ekers puts it, Zwicky was more about hand-waving and proposing grand ideas based largely on intuition, "so he would not have found neutron stars interesting once it all became a more detailed analysis."[16] Zwicky died five years after neutron stars were discovered, on February 8, 1974, in Pasadena, California, and is buried in a small cemetery in Mollis, Switzerland. I add a line to my bucket list to go there one day and pay my respects.

While the Swiss eccentric couldn't be bothered with neutron stars now that they had been discovered, other scientists piled on the bandwagon, turning their telescopes in the direction of individual pulsars. They were keen to observe not only the neutron stars themselves, but also—once again based on Zwicky's theories—the nebulas associated with them, which he had posited as being remnants of these stellar deaths. One of the telescopes that researchers used for this purpose was the Molonglo Radio Observatory. Located about an hour's drive from the Australian capital, Canberra, and some four hours' drive from Parkes, Molonglo is a weird construction, with two perpendicular arms each a mile long—radio antennas called the Mills Cross. The arms are in the form of half-cylinders, and the antennas employ a parabolic cross-section shape to focus the signal. It's possible to rotate one of them, moving the antenna up or down by steering around its axis.

The telescope became fully operational in 1967—just in time for the very first pulsar, LGM-1, to make its entrance. Molonglo was built to observe radio sources, so the pulsars were right up its street. The relatively narrow diameter of the cylinder, compared to larger single dishes, meant that the telescope scanned a much wider swath of the sky passing overhead and hence could search for pulsars faster. Straight away, it began picking up pulsars—so many, in fact, that more than half of all pulsars and supernova remnants detected in the first two decades of pulsar astronomy were spotted at Molonglo. Since the 1960s, the telescope has been through a number of upgrades and is now called the Molonglo Observatory Synthesis Telescope (MOST). I visited the observatory with Matthew Bailes on a particularly windy and gloomy day in February 2019, with a steady drizzle spoiling any hopes for an Australian summer's day. A telescope operator on shift greeted us near the small house serving as a control room and data center, storing data from decades of celestial detections. The operator led us to the telescope, which looks much more like a gigantic agricultural irrigation tool than a precise astrophysical instrument. A flock of sheep belonging to a nearby farmer only reinforced the association. Spooked by our presence, they quickly took off, running under the north-south arm in ever-stronger rain.

It is this telescope that, on Friday October 4, 1968, proved Zwicky right once again. Astronomer Michael Large was at Molonglo's control room, a small house, observing newly found pulsars. He was staring at the data pouring out of the chart recorder when the two pens started scribbling erratically. Large knew he had detected a faraway space body. "When he saw it, he got a message through to me immediately," says astronomer Alan Vaughan, then working at Molonglo as part of his PhD thesis; Large was Vaughan's PhD supervisor. Vaughan's job was to improve the telescope's sensitivity. That Friday though, he was at a Christian conference, and hard to reach. Not until Large called Vaughan's home and Alan's mother gave him the phone number of the conference center did the two men connect. Large was very excited—and Vaughan promptly got on a train to get to Molonglo. He got there just in time for the next transit of the newly found pulsar, twenty-four hours after it was first seen.[17]

"We made a whole lot of preliminary measurements, and it was remarkable—it was very fast compared to all the other pulsars known at the time," says Vaughan.[18] Indeed—the neutron star was pulsing about eleven times per second, whereas the other six known then (two detected by Molonglo in previous months, the rest by Bell) pulsed once every second or so.

Vaughan and Large rushed with their data to the University of Sydney. There, they mentioned the coordinates to their colleague—and the scientist behind the design of the telescope, Bernie Mills. "Bernie straight away knew that the source was at the location of a supernova remnant," Vaughan remembers.[19] The nebulous remains of a massive star that had died some fifty thousand years ago had a spinning, highly magnetized heart. Dubbed the Vela Pulsar, it became the first direct link between a neutron star and a supernova. The researchers rushed a paper to *Nature* and it was published within weeks of the discovery, on October 26, 1968. Excited as they were, they didn't throw a party—instead, they got on with their work searching for more pulsars, says Vaughan. Nearly a decade later, in 1977, astronomers using the Anglo Australian Telescope would also observe the Vela Pulsar—for the first time, in optical light.

On exactly the same day that *Nature* received the paper about the Vela Pulsar, two other astronomers, David Staelin and Edward Reifenstein, were using the Green Bank Telescope in the United States, about two hours' drive away from Charlottesville, West Virginia. They spotted peculiar pulses streaming from the Crab Nebula—another known supernova remnant in the constellation Taurus. As mentioned earlier, Chinese astronomers had recorded this bright supernova in 1054. And in 1850 astronomer William Parsons drew its remnant after spending many nights monitoring it with his thirty-six-inch telescope. Because his drawing resembled a crab, the remnant became known as the Crab Nebula.

One month after the Green Bank Telescope's observations, in November, Cornell graduate student Richard Lovelace detected that the radio signal coming from the Crab Nebula arrived at regular intervals, every 33 milliseconds. He was using the Arecibo Telescope in Puerto

Rico—a huge dish built into a natural sinkhole in the ground, which was later made famous both by the film *Contact* (which was also shot at the Very Large Array in New Mexico) and the James Bond movie *GoldenEye*. The results from Green Bank and Arecibo validated the prediction made by astronomer Franco Pacini, who months earlier had published a paper arguing that there would be a pulsar inside the Crab supernova remnant.[20]

So now there were two pulsars shining brightly inside supernova remnants. The pieces of the jigsaw puzzle were falling into place, confirming all the hand-wavy predictions made by Fritz Zwicky in 1933. Neutron stars—hyper-dense, tiny, rapidly rotating, and highly magnetized—had become very real.

SUPERNOVAS ARE RARE; they happen only about twice a century in a galaxy the size of the Milky Way; our galaxy, home to some three hundred billion stars, is already overdue. The most recent known supernova in our galaxy happened near the center of the Milky Way around 1870, although it was not visible because it was shrouded by dust; astronomers can now observe it as a radio and x-ray remnant. Because there are at least two trillion galaxies in the observable Universe, roughly ten stars go supernova every second somewhere in the vastness of the cosmos.[21]

There doesn't seem to be any regularity as to when and where one might appear, though, which means that they are notoriously difficult to spot. Astronomers use wide sky surveys performed by robotic telescopes that constantly monitor all the flashes they can detect across a range of wavelengths. Algorithms sift through the data to identify promising supernova candidates and try to quickly get time on a powerful telescope to study the burst's spectrum of light—its astrophysical fingerprint. Matteo Cantiello, an astrophysicist at Princeton University, calls such analysis stellar forensics: we know that a star has died, and now we want to find out exactly why and how. Observing the last few hours of its life, if possible, can give astronomers crucial insights into supernovas—

including figuring out whether a dying star showed any signs of imminent detonation during the months, weeks, or days before blowing up.[22]

On February 23, 1987, astronomers got a rare glimpse of a supernova in action. The night shift at the Las Campanas Observatory in Chile had just started. Astronomy there was still done the old-school way, with scientists on location right at the telescope, not in a comfortable office hundreds of miles away at the other end of a high-bandwidth internet connection. Observers Ian Shelton and Oscar Duhalde were on duty, tracking the sky. Suddenly, Shelton spotted a flash of light coming from the direction of a dwarf galaxy not far from the Milky Way, the Large Magellanic Cloud (LMC), some 168,000 light-years from Earth. It was bright enough to be visible even to the naked eye. He quickly recorded the observation (learning later that another astronomer, in New Zealand, had also seen the light). The scientists checked the catalog and saw that there used to be a massive star at the source of the flash—a star that had suddenly vanished. The blue supergiant Sanduleak −69° 202 that used to shine on the outskirts of the Tarantula Nebula in the LMC, was no more—it had exploded in a Type II core-collapse supernova that later became known SN 1987A. For Earthbound viewers, it was the nearest supernova that had been seen for more than four hundred years.

Three hours before the astronomers in Chile and New Zealand spotted the flash in optical light, three neutrino detectors—Kamiokande II in Japan, Irvine-Michigan-Brookhaven (IMB) in Ohio on the shores of Lake Erie, and Baksan in Russia—caught in total about two dozen neutrinos. That was many more neutrinos than any of the three observatories had recorded at any point previously. Until then, the neutrino mechanism of a supernova had been merely a theory. Now there was observational proof.[23]

But proving conclusively that supernovas are powered by neutrinos is much more difficult. Researchers have been running 3D computer simulations of the collapsing core of a star and its neutrino-driven mechanism, a very expensive and time-consuming computational process. Running just one calculation can take many months. It doesn't help that it's very difficult to observe the mechanism that causes a supernova explosion, "because all you see is when the shockwave hits the surface," says Stephen Smartt at Queen's University Belfast. The light that is detected is determined by the radius of the star and the amount of material around the star—features that are not particularly useful for understanding the explosion mechanism. "You have to infer that from the kinetic energy of the material that you measure, the movement of the material, how much energy is in the shock and in the exploding star," says Smartt. "It's difficult theoretically, and it's difficult to probe right into the center of the explosion."[24]

Researchers have now been studying the evolution of SN 1987A for more than a quarter of a century. They've seen its transition from the nebular phase to a supernova remnant in real time. One big mystery has been its missing neutron star, which theory suggests should remain at its heart, given the size of the progenitor star, which had about twenty solar masses. The neutrino observations suggest that a compact object did form at the original star's core. But so far astronomers have found nothing. One possible explanation is that there's simply too much debris floating around, hiding the core behind gas and very dense dust clouds. Or perhaps if there is a neutron star, its magnetic field is either too strong or too weak to form a normal pulsar. Alternatively, perhaps too much debris fell onto the young neutron star—so much that its mass increased to the point of no return, causing it to collapse further into a black hole.

No one had recorded any warning signs that the blue supergiant was going to die, despite researchers having observed the supernova very shortly after it exploded and having known the progenitor.[25] There are at least some stars, however, that seem to give us warning signs of their imminent demise. In 2013, astronomers found such a star. Typically, to find new interesting objects, a robotic telescope scans the night skies and human monitors—astronomers on shift—catch and put aside for follow-up observations the most interesting and relevant supernova candidates as they stream in live from the telescope. It means that events are spotted and can be observed as quickly as possible. So when a red supergiant died in a magnificent display of fireworks, a robotic sky survey called the Palomar Transient Factory in Southern California was on duty—and just three hours after the supernova's first light reached Earth, human monitors in Israel caught it. The star was in a relatively nearby galaxy and it had not been detected the previous night, which was intriguing.

As it happened, one team had already pre-booked time on the Keck Telescope in Hawaii for some other observations that night. Ofer Yaron, an astrophysicist at the Weizmann Institute of Science in Israel, alerted Dan Perley, a researcher at the California Institute of Technology who was observing with Keck. Perley immediately asked a telescope operator on a mountain in Hawaii to point Keck at the flash, while also quickly lining up the space-based Swift telescope to take data in x-ray and ultraviolet light.

Perley managed to take a sequence of four spectra, collecting light from the aperture of the telescope to a spectrograph, using slit spectroscopy—framing the object he was observing inside a slit of a certain width to get as much light as he needed and without getting more light from nearby objects. "These are still the earliest spectra ever taken of a supernova explosion, as of today—2020," says Yaron.[26]

Until then, astronomers had thought it was impossible to know if a star was going to explode within ten thousand years. But Yaron's team found that in the future, it may be possible to observe warning signs of a giant star's imminent demise a few years or perhaps even months before it goes supernova, with the star ejecting material at a faster and faster rate—just like a volcano sometimes gives indications that it's about to erupt through the tremors caused by rapidly rising magma.[27]

When Yaron and his colleagues analyzed the spectrum from the supernova they were observing, dubbed SN 2013fs, they found that there was a dense shell of gas around the dying, exploding star—what's known as circumstellar material. Was this gas there for hundreds of years before the explosion, or could it have been an early sign that the star was about to erupt? The theory of core-collapse supernova suggests that looking at a star will offer no clues to whether its innards are already violently collapsing in on themselves before the final detonation. The mantle, the outer part of the star, stays eerily quiet until the final moment when the core rebounds against itself—and the mantle bursts, all in a matter of seconds. But even before Yaron's observations, other research teams had been suggesting that dying stars were indeed shedding some of their outer layers of gas before going supernova.

In one piece of research, scientists analyzed images taken by a robotic supernova survey of sixteen supernovas and found that five of them did show minor flashes before the actual explosion. Yaron's paper shows that the progenitor star his group analyzed did indeed shed much more gas during its last moments than at any time during the preceding years. The pre-supernova star underwent a number of eruptive mass loss episodes. This gas disappeared just five days after the explosion, as the shock wave from the supernova spread into interstellar space.

Not everyone agrees with this analysis. Other astronomers think that because red supergiants are surrounded by plenty of gas all their lives, it's possible the circumstellar material was there for a long, long time. "It could be that the star's cloud was sitting there for millions of years," says Norbert Langer, an astronomer at Bonn University. Take Betelgeuse, for instance, in the constellation Orion, which at more than six hundred light-years is fairly close to Earth and is clearly shrouded in gas that probably has been there for thousands of years, if not more. To solve the conundrum, we would need to measure how fast the gas is speeding away from the exploding star. If quickly—then it has just exploded. In the case of SN 2013fs, it was possible only to get the upper limit on the velocity, but Yaron thinks there are good indications that the gas was moving much faster than typical wind velocities of a red supergiant.

So how likely is it that the outer core "knows" that the star is about to explode? Stars are so large and dense that it would take thousands of years for any information to flow from the core to the surface. But information doesn't have to be carried by photons, which have a hard time passing through the thick innards of a star. Instead, it can be carried by shock waves—and with this energy, create massive bubbling on the surface. In 2017, Caltech astrophysicist Jim Fuller published a paper that suggested that information on a star's inner workings could be carried to the surface by sound waves. He compared the process to a pot of boiling water—if the boiling is intense, one can hear it, because the process excites sound waves in the air. It is conceivable that the core of a star experiences "boiling" just before it goes supernova, which would generate hugely energetic sound waves that might trigger outbursts on the surface right before the big boom.[28]

Yaron and Fuller are not the only ones who think there may be observable warning signs in a star before it goes supernova. Anna Ho, in addition to her research on the Cow, has been studying SN 2018gep, a Type Ib supernova found by the Zwicky Transient Facility. Her team started observing the supernova within hours after the light from the explosion arrived on Earth, and obtained nine spectra within the first few days of the outburst—the earliest spectra of this kind of supernova. Unlike Yaron's data, this event is a so-called stripped envelope core-collapse supernova—where the parent star loses its hydrogen outer layers before the blast. But just as with Yaron's event, Ho's analysis shows that the star showed warning signs of imminent explosion. This is the first definitive detection of progenitor eruptions in this type of supernova, and it could mean that such warning signs are much more common for different kinds of massive stars, not just those that explode as a Type II supernova.[29]

While Ho's and Yaron's data on the moment of the death directly probe the composition of the progenitor star, they can offer only a limited number of insights into the supernova process. Most of the energy from the collapse is driven out by neutrinos, and for Yaron, the question now is whether enough of this energy can be coupled to the gas—to the material of the star and the outer envelope of the star, thus driving the supernova ejecta.

To peer inside a supernova, astronomers look into its nebular phase, some two hundred days after the burst. They also observe the supernova remnant that forms decades later. That's when the ejecta has moved beyond the explosion to interstellar space, where they can be seen relatively easily. In a way, the explosion has turned the original star inside out. It's in these stellar remains that scientists can find the heavy elements that lead to the creation of planets, other stars, and ultimately, life.

Taking spectra of the later stages of a supernova evolution makes it possible for astronomers to estimate the mass and the abundance of each of the components that were created and ejected during the supernova; they also can estimate the overall mass of the progenitor star and its composition.[30]

After this nebular phase comes the remnant phase—and that's exactly what we now see forming in real time with SN 1987A. While it's possible that there were early warning signs for SN 2013fs, this doesn't seem to be the case for SN 1987A—despite both being Type II core-collapse supernovas. The available data about the progenitor of SN 1987A don't seem to point at any warning signs like eruptions on the surface: it was a pretty well-behaved star during the century before it exploded.

While "warning signs" data are valuable, the key now is to apply similar analyses more often, with many more supernovas, by getting the spectra information early, that is, before the ejecta overruns any circumstellar material, erasing all the forensic information about the actual death of the supernova. And that's where the Cow can help.

SUPERNOVAS ARE USUALLY OBSERVED in radio wavelengths: as the stellar material expands away from the explosion, the energy carried outward starts at extremely short wavelengths, but very quickly produces waves that are longer and longer. It's very rare that astronomers resort to millimeter telescopes to study cosmic explosions, though—because to see the signature in the millimeter waves, one has to be very, very fast. "So by the time you observe an explosion, a signature has already moved to these longer radio and centimeter-long wavelengths," Ho says.[31]

But her team was fast enough—and the supernova turned out to be spectacularly bright in millimeter wavelengths. The result: the first-ever observation of a supernova increasing its brightness in the millimeter wave range—a crucial piece of data to understand these mysterious explosions. The short-wavelength radiation from the Cow continued for weeks, enabling the team to observe it for eighty days.

"The strange, mysterious thing about the Cow is that if you just look at the optical spectra, it really doesn't look like a supernova," says Ho. To figure out whether this truly was a stellar explosion, you have to look at other pieces of evidence. Her team was able to study the gas and dust around the explosion, and found that the characteristics of environment were very similar to the aftermath of a stellar explosion. In particular, Ho saw that the density of the gas and dust was very high—the kind of density that you might expect of a star before it died, when it was violently shedding material—exactly what massive stars are thought to do before they explode. So perhaps before this star died, it erupted and shed a lot of material, and when it finally exploded, it interacted with that material. The ultra-short wavelength radiation lingering for months also suggests that there wasn't a single release of energy in a blast but rather some sort of ongoing energy production like that created by a central engine—either a rapidly rotating, highly magnetized neutron star, or a newly born black hole accreting matter.[32]

It's not just Ho's team who believe that they were watching the birth of a neutron star as it happened. Another group, led by Raffaella Margutti, an astrophysicist at Northwestern University, came to the same conclusion. Margutti's team studied the Cow for its x-ray emissions, using NuSTAR and the European Space Agency's INTEGRAL space telescopes, plus the radio wave spectrum with the National Radio Astronomy Observatory's Very Large Array in New Mexico. They too think that the transient was being reheated from the inside—meaning that something was powering it from within.[33]

But to confirm whether or not there is indeed a neutron star hiding inside the Cow, astronomers will have to wait years until the debris—the gas and dust around the center—has dispersed into interstellar space. "We see stars and separately we see explosions, and again, separately, we see neutron stars. But it's very difficult to link these different steps together," says Ho. "So if we were actually able to see that there is a neutron star left over from the Cow, that would be one of the only times we've ever actually seen an explosion producing a neutron star. It's actually witnessing this moment when a neutron star is born—it would be extremely exciting."[34]

If it were confirmed, future observations could tell us more about the kinds of stars that result in neutron stars. At the moment, we are able to study neutron stars because we see them in our own galaxy. We know—or we think we do—that stars have to be of a certain mass to explode. But what was the star that became a neutron star? What was its evolution? What was it doing during its life? What were the final stages of its life like? Was it losing mass, and if so, was it doing so in a constant way or in a violent, eruptive way? We really have no idea. So if there is a neutron star in the Cow, this zombie star could tell us a lot about living stars. But for now, we have to wait for this neutron star, assuming it's there, to reveal itself to us.

As I leave ALMA, I know that one day I will come back. This mesmerizing high-altitude Chajnantor plateau, the array of otherworldly-looking antennas, and all the other telescopes dotted across this endless, rocky, arid, but somehow astonishingly lovely place are a temptation I won't be able to resist. So long, Atacama.

DEEPER DIVE: Pulsar Kicks

During the past half century, astronomers have detected nearly three thousand radio pulsars and have observed several dozen supernova remnants in the Milky Way. But we still don't know much about the fine line between a stellar death and a neutron star's birth. For instance, we know that stars move—they orbit the center of their home galaxy, usually moving at between 65 to 100 km (40 to 62 miles) per second. But we don't know why newly born neutron stars are often kicked out of their supernova remnant, moving at 200 to 500 km (124 to 310 miles) per second in the direction opposite to the supernova ejecta, a much higher speed than the original star used to have.

Scientists call it the pulsar kick. The kicks are probably due to a supernova exploding not as a perfect symmetrical sphere but asymmetrically, says Thomas Janka, an astronomer at Max Planck Institute for Astrophysics in Germany—just like cigarette smoke that's rising smoothly can suddenly develop vortex flows.[35]

In March 2019, astronomers using NASA's Fermi Gamma-Ray Space Telescope managed to track a pulsar called J0002 + 6126, some 6,500 light-years away from us, which was flung from where its star exploded and is zooming through space at four million kilometers an hour. The pulsar is moving almost twice as fast as the average pulsar and is leaving a glowing tail thirteen light-years long in its wake. If a spaceship were flying at this speed to the Moon, it would arrive in a mere six minutes (whether it could brake in time is another matter).

Understanding the asymmetry of the supernova would not only help us to analyze the direction and power of the kick, but could also shed light on a much bigger mystery: what's the actual mechanism powering a supernova? The most broadly accepted theory right now suggests a neutrino-driven mechanism—where neutrinos are produced in huge numbers during the core collapse, then stream out asymmetrically. "Detailed computer models of such explosions should explain not just neutron star kicks, but also the many observational hints which show that supernovas eject their gas highly asymmetrically," says Janka. Three-dimensional computer models of supernovas, such as the precise model reproducing the geometry of the four-hundred-year-old supernova remnant of Cassiopeia A, have supported observations like pulsar kicks, which is a major step forward for theorists and astronomers alike. "Today's 3D supernova models are found to yield neutron star kicks of typically several 100 km / s with extreme cases up to 1000 km / s and possibly more," says Janka. "This is consistent with the measured proper motions of young neutron stars."

The models have also been able to explain the distribution of iron around the supernova, and of recently measured radioactive titanium, which were observed in detail by the NuSTAR satellite, a space-based x-ray telescope that observes x-ray radiation from various objects, including neutron stars.

While neutron stars receive their kick from the asymmetric supernova explosion, the pulsar's spin is thought to result mainly from the rotation of the progenitor star, because its angular momentum is conserved when its core collapses to become a neutron star. So if these assumptions are correct, then the rotational speed and the motion through space are powered by two different mechanisms. To measure the velocity of the motion through space, one has to know the distance to the neutron star, which is not always easy to determine. Once that's done, astronomers are able to measure the velocity of the motion only in relation to the plane of our sky; the radial component of the velocity, toward or away from us, cannot be measured. To make such measurements, astronomers compare direct observations of the changing celestial position of the pulsar, or through radio measurements of its radiation.

Still, some pulsars, like the Crab and Vela pulsars, seem to stay inside their nebula homes. Janka says that not all neutron stars receive high kick velocities; it all probably depends on the explosion energy and thus original mass of the parent star. The more massive the progenitor star, the more powerful the explosion will be, hence the stronger the kick, the theory goes. "A lot of neutron stars have velocities of less than 200 km / s, and stellar evolution theory suggests that a separate population of low-velocity neutron stars might also exist, those with velocities of only some 10 km / s." Indeed—this population could explain why we observe quite a number of neutron stars in globular clusters, big groups of stars sticking together. The velocity that stars would need to escape the cluster is quite low, about 30 to 50 km per second, so neutron stars observed there had to be born with low velocities.

Another reason why some neutron stars have left their remnant nebula while others have not is thought to be connected to the age of a supernova remnant. Initially, the remnant gas seems to expand into circumstellar space at speeds of several thousand km per second, so about ten times faster than the average speed of a neutron star. Only some ten thousand years later would the supernova ejecta gas gradually slow as it interacts with the interstellar medium—the surrounding gas and dust. But the neutron star doesn't slow down; it continues to move with its initial speed. So at some point, possibly many thousands of years after the supernova blew up, the neutron star would be able to leave the gaseous remnant. When exactly this happens depends on the neutron star's velocity and the density of the interstellar medium with which the gaseous supernova ejecta interacts, says Janka.

DEEPER DIVE: The Death of Massive Stars

Stars die differently, depending on their mass. Some go out rather quietly, others—with a bang. Stars that are about eight times less massive than our Sun live for hundreds of millions to billions of years—the longer the less the mass the star has. They then shed off their outer layers in planetary nebula while their stellar core shrinks to a white dwarf, its compact remnant, that will stick around in the vastness of the cosmos for eternity.

A very different fate, however, awaits more massive stars, like the red and blue supergiants that weigh as much as eight to one hundred times our Sun. At the time of their death, their cores, in most cases made of iron, will exceed the limit of roughly 1.4 solar masses (the real number can be between about 1.3 and more than two solar masses). They have enough nuclear fuel to burn for millions of years, shedding matter into interstellar space as they do so. Their cores are kept from collapsing by the heat produced when the nuclei of lighter elements get fused into heavier ones, producing all known elements—from carbon and oxygen to iron—that the star burns through in turn. The elements are kept locked inside these stars, at least for the time being. Nearly massless elementary particles called neutrinos are also created.

In the final years of the star's life, this nuclear fusion and production of elements accelerates enormously, with temperature and pressure going through the roof. While the hydrogen in such a star will burn for many millions of years, the helium will last for just one million years or so. When it's gone, carbon fusion starts, which lasts for about one thousand years. That's followed by burning of the remaining oxygen, which lasts for a mere few weeks. Heavier and heavier elements undergo the same fate, ever faster: after oxygen, neon burns, then magnesium, and finally, at the very last stage, silicon—just before the core is turned into iron, which probably takes only a single day.

And that's when a star's problems really start. Unlike all previous elements, iron doesn't create energy during fusion; it demands it. At that point, however, a star's energy production has already ceased, and the core starts to contract, since gravitational energy is the only source of energy available. As the core's mass—now made from iron—reaches the approximate 1.4 solar masses threshold, suddenly there's not enough resistance in the core to counteract the crushing force of gravity. In a fraction of a second, the outer edges of the core collapse, moving inward at speeds of more than seventy thousand meters per second (roughly 23 percent of the speed of light). The collapse crushes protons and electrons together to produce neutrons, which in turn generate neutrinos that zoom out of the collapsing star. The infalling material rebounds off the iron core, which now reaches a density of an astonishing 400 million million grams per cubic centimeter—and in the process gives birth to a neutron star (or a black hole, if the star was very, very massive).

Now a hydrodynamical shockwave forms around the core, heated by the energy of the neutrinos racing away in all directions; they fly in straight lines and will arrive at Earth before the light of the supernova, because the actual explosion has yet to happen. As the shockwave moves beyond the core to the surface of the star, which can take a couple of hours, the star's inner shells reach ultra-high temperatures, triggering the formation of elements on the periodic table higher than iron—the iron-group nuclei, such as radioactive nickel-56, and intermediate-mass elements between the iron group and silicon, for instance radioactive titanium-44 (most heavier elements such as gold, platinum, silver and uranium are thought to come from neutron-star mergers). As the shockwave reaches the outer layers, or mantle, of the star, it's blown away in a huge blast that reaches speeds of more than 50 million km (31 million miles) an hour. It's this final blast that announces the star's demise in a splendid supernova display that can outshine an entire galaxy.

This type of supernova is called a core-collapse supernova. There are many different types of core-collapse, but the most common ones are Type Ib, Type Ic, and Type II. In all three instances, the progenitor stars originally had two layers, hydrogen on the outside and helium just underneath. Stars that shed both layers during their lives explode as Type Ic; those that have the helium layer still attached are Type Ib; if both layers are still there during the core collapse, then scientists call it a Type II. Yet how exactly a neutron star is born out of these core-collapse scenarios is still unclear, despite decades of research, modeling, and observations.

In recent years, scientists have also been puzzled by a completely different type of core-collapse supernova: the so-called superluminous supernova. These amazingly energetic transients are rare and about fifty times as bright as their ordinary core-collapse cousins. We have no solid explanation for what their nature is and why they are so bright, but theories abound. One is that they could be powered by highly magnetized, rapidly spinning neutron stars that form when an ordinary core collapses and transfers its energy into the supernova ejecta—the star's shreds—superheating them to the point of visible light. In coming years, surveys such as the Large Synoptic Survey Telescope are set to find millions of new supernovas. Analyzing the results will be tremendously useful for understanding how and why giant stars actually explode—and astronomers can't wait to get their hands on this promising data to come.[36]

• 4 •

Zombies and Starquakes

CHIA MIN TAN WAS ON VACATION, at his home in Malaysia. It was August 7, 2017, a quiet summer day, and he had just had lunch with his mother and brother. While the rest of the family sat down to watch some TV, Tan decided to sift through data from a pulsar survey he had been working on before his trip home from the University of Manchester. A graduate student exactly one year, ten months, twelve days, and twenty hours into his PhD, he was keen to complete his thesis as soon as possible, on a long road toward his dream of becoming a full professor.

Tan was looking at data taken during the LOFAR Tied-Array All-Sky Survey (LOTAAS) with a very unusual and fairly new telescope in the Netherlands called LOFAR. Short for Low Frequency Array, it's the world's largest and most sensitive low-frequency telescope, operating at observing frequencies in the 10 to 240 MHz range—the lowest frequencies we can detect from Earth.[1] The signals he was staring at suggested that there were several new pulsars in the location they were coming from, with different periods (the time it takes a neutron star to make one full revolution around its axis of rotation), all at the same distance from Earth and in the same patch of sky. Intrigued, Tan did further analysis and realized that they were all from the same pulsar but detected at different fractions of its incredibly long period of 23.5 seconds. He was stunned, since he knew that this pulsar was slower than any other known to exist. Tan immediately sent an email to the LOTAAS collaboration to announce the discovery, and fellow astronomers straight away set up a meeting to discuss the next steps—but promised they would wait until he got back from vacation.

With his return flight just a couple of days later, Tan kept thinking about his find. The pulsar, now called PSR J0250+5854, was so slow that on the diagram astronomers use to plot the pulsar population based on each star's speed of spin (called the p-p-dot diagram), it landed beyond the so-called "death line." This theoretical line denotes the boundary beyond which pulsars are assumed to die—when they spin so slowly that they no longer emit radio waves we can detect. When that happens, there is no way for us to spot the neutron star unless it revives (which can happen in very special cases). Tan's pulsar, however, pushed the death line way beyond where scientists had expected it to be.

The discovery was quite surreal for Tan, because he never expected to find a new pulsar with such significance, especially during his PhD work. Discovering it at home, however, somewhat dampened the excitement, because there were no colleagues around to share it with. He told his family that he had found something unusual, but it was hard to explain the significance to them. Tan couldn't wait to get back to Manchester.

There were a few panicked moments when his flight was delayed by half a day. In the end, Tan managed to get to his university just in time for the meeting scheduled to discuss the new pulsar, carrying his suitcase right into the meeting room. His colleagues were waiting for him, and together they discussed the upcoming scientific paper and the significance of the discovery. The find showed that there are likely to be more slow-spinning pulsars out there that have not yet been detected. "Finding them will give us a better idea on the population of pulsars and neutron stars in the galaxy," Tan says, and will also help astronomers estimate how quickly neutron stars have been forming during the evolution of the Universe. Another peculiarity of this pulsar is that while it is spinning very slowly, the measured rate of slowdown of the spin suggests that it has a magnetic field that is strong relative to most pulsars.[2]

After being found by LOFAR, the pulsar was later also picked up by other telescopes sensitive to radio frequencies of 300 to 400 MHz. And like many astronomers nowadays, Tan has not yet visited the telescope he is using; there's no direct need—all data are taken and analyzed remotely. I, however, want to visit LOFAR, to see this very unusual telescope with a completely different design from any other pulsar-detecting machine on Earth.

AFTER A QUICK TWO-HOUR train ride from Amsterdam, I get off at the town of Hoogeveen in the northeast of the Netherlands. It's early May. The scene is postcard pretty; there are picturesque Dutch houses with ornate windows right next to the train station.

Frank Nuijens, the smiling science communicator at the Netherlands Institute for Radio Astronomy (ASTRON) that manages LOFAR, invites me to his car. Half an hour later we park it in a forest, not far from the village of Dwingeloo. We are at ASTRON's headquarters. It's also home to the Dwingeloo Radio Observatory, a single-dish, twenty-five-meter telescope that when completed in 1956, was the world's largest. The distinction didn't last long—the seventy-six-meter Lovell in Jodrell Bank overtook it the following year. I visit Dwingeloo's control room first. Since 2000, this dish has been used by amateur astronomers and students, and one of its more unusual projects is the Earth–Moon–Earth communication or "moonbounce." When astronomers send radio signals to the Moon, the waves bounce off the lunar surface and are detected by another telescope elsewhere on Earth—effectively allowing people to communicate via our natural satellite.[3]

As I climb up a few steps to get into the control tower, the tower itself starts rotating; unlike Parkes, where the tower is stable and only the dish moves around, here everything moves when the telescope is swiveled to point at a different part of the sky. It's a slow but disconcerting rotation. Pine trees sail past the tiny window, which makes my brain think that I'm on a ship in the open sea. Fighting back motion sickness, I smile at Nuijens and the telescope operator, who excitedly gestures at his computer screen. "Here, look, that's a pulsar we are pointing at right now," he says. The signal on the screen goes up and down, with regular peaks just like a human heartbeat. The operator cranks up the volume. I hear "Blip . . . blip . . . blip . . ." every few seconds. I am listening to a pulsar in real time. A pulsar somewhere far away in our Milky Way galaxy. Crazy.

Forty minutes later, and we're finally on our way to the place I've come to visit: LOFAR. After an hour or so on regular roads, we drive into real countryside, with occasional farms whooshing past. Unexpectedly, Nuijens stops the car. It's a dirt road, endless fields both ways, somewhere between the villages of Exloo and Buinen. "We are here," he says, but I struggle to see anything that looks like a telescope. The field is swampy and my feet get wet in seconds; a goose flies idly by and honks, eyeing us suspiciously. There's no magnificent antenna that can be seen for miles, like at Parkes in Australia; there is no collection of smaller dishes that from afar look like a cluster of mushrooms at ALMA in Chile's Atacama Desert or at MeerKAT in South Africa. After walking for a couple of minutes, we come to an oddly flat structure spread out amid the wetlands. Dozens of swans and ducks swim in a nearby creek. The structure looks like a collection of solar panels that have fallen flat. Next to them, weird metal poles stick out from the ground; they are a bit shorter than me and resemble old antennas for analog TVs.

This is the core of LOFAR. Once agricultural land, the entire area has been transformed into a nature reserve of four hundred hectares (1.5 square miles), following an agreement between the Netherlands Institute for Radio Astronomy and the local government. The astronomers wanted a remote area with minimum radio interference; in return, they promised to create a thriving wildlife sanctuary. Work on the observatory started in 2006. The epicenter is a collection of twenty-four core stations placed within a 3.2 km (two-mile) diameter in this swampy field. There are also fourteen other stations spread out across the Netherlands over some 96.5 km (sixty miles), as well as fourteen international stations in Germany, France, the United Kingdom, Ireland, Sweden, Poland, and Latvia.

What makes LOFAR unique is its ability to detect very low frequency radio waves, arriving in the range between 10 and 240 MHz—overlapping the FM radio band, which ranges from 87.5 to 108 MHz. Before the data arrive at the computers of researchers like Tan, they have been processed by ASTRON's supercomputer hosted at the University of Groningen, which takes in the signals from LOFAR in real time. Capturing data with this observatory is very different from doing so with other telescopes. Instead of collecting radio waves from a specific patch of sky and focusing the signals onto a receiver, LOFAR combines signals from thousands of antennas in clusters across several countries, and uses a technique called interferometry to get them to make sense together as one gigantic virtual telescope, one with a total effective collecting area of about 300,000 square meters (186.4 square miles). That's also how arrays of parabolic dishes work, except that LOFAR does not have any moving parts. Every one of its antennas sees the whole sky at any time, and to "point" LOFAR toward a particular patch of sky, the supercomputer calculates the difference in arrival times of a radio signal coming from the patch of sky across various antennas, then adds all the signals by correcting for the time difference; at the end all signals are aligned to map the sky.[4] "This is very useful as we can point the telescope to hundreds of different directions at the same time, to survey a large part of the sky in one go," Tan told me before my journey.[5]

The very low frequencies captured by LOFAR have long been ignored by single dish observatories. That's because radio waves are very long—and the lower the frequency, the longer the wavelength, with 10 MHz corresponding to a wave 30 meters (98 ft) long. That's why it helps when the collecting area is extremely large; it gives scientists a good resolution—and LOFAR has dramatically broadened the ability to detect radio sources across the radio spectrum. So far, this telescope has found nearly one hundred new pulsars.[6]

As I look around, I realize that the creek with the ducks and swans winds nearly all around us. We are standing on a circular "island" some 320 meters (1,050 ft) in diameter, called a Superterp—the heart of the core sporting six receiving stations. In the Dutch dialect spoken in this part of the Netherlands, a terp means an artificial elevated site that provides protection in areas prone to flooding. Indeed, with water all around, I can see the need. From high up, in the images I look at later, the Superterp looks as if some alien flying saucer has left an imprint after taking off. Each station inside the circle is made from two flat structures with twenty-four tiles, each five meters (16.4 ft) square, covered by black tarpaulin. Frank carefully lifts the cover on a corner of one of the tiles, revealing a white Styrofoam frame underneath. This hides a sensitive "bow-tie" antenna in a honeycomb structure—a setup that allows each antenna to "see" the entire sky at once and point the telescope in different directions simultaneously. Another goose flies past us and rain starts to drizzle; carefully, we cover the antenna and walk back to the car. I feel at once both awed and underwhelmed: this unspectacular telescope looks painfully boring, but it has just detected the slowest pulsar ever.

THE PULSAR THAT CHIA MIN TAN FOUND wasn't always slow. When it emerged from its supernova explosion, it was young and energetic and was spinning much, much faster. That's because when the core of a pulsar's parent star collapses, with protons and electrons crushed together and morphing into neutrons, the collapse ends only if the pressure of the neutrons squeezed ever so tightly into a tiny space counterbalances the core's enormous gravity. Such a newborn neutron star will spin like crazy—inheriting its rotation from the original star. All stars spin; our Sun does too, rotating around its axis about once every thirty days. But when a massive star at the end of its days shrinks rapidly during the gravitational collapse, it gets a huge rotational velocity kick—similar to when a figure skater suddenly pulls in her outstretched arms—making it spin much, much faster. In other words, the star transfers its angular momentum (spin) to its shrunken core, giving it a huge speed boost.

We can detect these rotating stellar leftover cores thanks to the radiation they send into space. The exact emission mechanisms are still a mystery, but scientists do have a few ideas. While all neutron stars are thought to have the same internal structure, there seem to be three different ways in which they emit observable radiation. Some do so simply because of their rotation and others by syphoning matter from their companion star. Finally, some neutron stars are so highly magnetized that sometimes their magnetic fields twist and turn enough to crack their surface, resulting in starquakes that spit out powerful flares of particles. All three mechanisms are fascinating in their own way—and the more we observe them, the more we understand what makes neutron stars tick.[7]

Rotation-Powered Pulsars: Spinning Galactic Wheels

When Jocelyn Bell found her four LGM pulsars, they were scattered all over the sky, spinning away in cosmic solitude. They were so-called rotation-powered pulsars, which produce radiation beamed along their magnetic fields and as a result lose energy as they spin. So far we've detected more than 2,700 of them, mostly in the Milky Way—their emissions are too weak for our telescopes to detect those much farther away. Tan's ultra-slow pulsar is one of these rotation-powered neutron stars. As it rotates, radiation is beamed by particles accelerating along its magnetic poles, and if this beam happens to cross the line of sight of our telescope, we register it as a pulse.[8]

Rotation-powered pulsars can be either standalone or part of a binary system—when two stars are gravitationally bound to each other and orbit a common center of mass. As they emit radiation, they lose energy, getting more and more tired—and hence rotate slower and slower, spinning down. Like Tan's slowest pulsar, they usually emit radio waves, although some also produce x-rays and gamma rays. A few rare beasts dubbed X-ray Dim Isolated Neutron Stars (X-DINs or XINS) seem to emit only in x-ray—unusual for a stand-alone pulsar. Only seven of these have been discovered, called "the Magnificent Seven." What they really are is a mystery. Perhaps we simply can't see their radio emissions because their jets are not pointing toward us. Or perhaps the emitted beam is extremely narrow, so the probability that it will cross a telescope's line of sight is very small. Imagine holding a laser pointer with its narrow beam and waving it around—the chance that the beam will ever shine directly into your eyes is very, very small.

Even before the discovery of the first pulsars, the Italian astronomer Franco Pacini had theorized that rotating neutron stars might emit radio waves. A lot of physics is happening, though, for them to be able to do so, and it all starts with the parent star, which transfers not only its spin, but also its magnetic flux (the component of the magnetic field perpendicular to the area it penetrates) to the offspring. That's how a neutron star gets its magnetic field. While pulsars have two magnetic poles—just like the bar magnets at physics lessons in school—they work quite differently. In a bar magnet, field lines leave the south pole and travel to the north pole where they will exit again, in continuous, endless closed loops.

When a neutron star spins, its magnetic field lines spin with it, or co-rotate—they are pinned to the surface like the lines on a spinning record. And just like a child on a merry-go-round, the farther out you are from the star, the faster you move. But this can't keep going forever—at some radius, the velocity would exceed the speed of light. At that point, the co-rotation with the field lines must break down—the magnetic field can no longer be rotating with the neutron star, because if it did, the field lines would be traveling faster than the speed of light, which is impossible.

This virtual boundary is called the light cylinder, beyond which co-rotation is not possible. And any magnetic field line that enters this region won't be able to close on itself to complete the loop. The radiation is beamed along directions that lie within these last closed field lines that end at the light cylinder. Every spinning magnet has a light cylinder—the Earth has it too. But because the Earth's magnetic field is fairly weak, ranging at the surface from 0.25 to 0.65 gauss, the edge of the cylinder is so far away that it doesn't matter. The neutron star field is so strong, however, that it does matter.

Those open field lines can accelerate particles, acting like a generator. They extract a cascade of particles from the neutron star's surface from around the area of the two magnetic poles. These high-energy particles, mostly electrons, zip along the open field lines in narrow, cone-shape beams that are not necessarily aligned with the pulsar's axis of rotation. As these highly energetic electrons interact with the magnetic field lines, they accelerate—and as a result produce radio waves that move in the same direction as the motion of the particles, out into the cosmos. As the beams sweep through space, if one of them happens to be directed toward Earth, we can home in on a pulsar. As mentioned earlier, just like a lighthouse at sea it emits a continuous beam of light, but as it spins around its axis, we see it as repeating flashes.

With time, as more and more particles are extracted from the neutron star's surface, the pulsar slows down. Deprived of energy, at some point it will be too slow for the magnetic field lines to pull more electrons off the surface, and the emission will cease. The Crab Pulsar is thought to have been rotating once every sixteen milliseconds when it was young, about 965 years ago; today, it spins once every thirty-three milliseconds and continues to slow down by about 1.3 milliseconds per century. The stronger its initial magnetic field, the faster a pulsar will slow down. Young radio pulsars have magnetic fields of around 10^{12} and 10^{13} gauss; for comparison, a typical fridge magnet delivers around a hundred gauss. Such pulsars will be active between ten and a hundred million years. This means that of all the neutron stars born during the 13.6 billion years of the Universe's existence, some 99 percent no longer emit radio beams—the light in the lighthouse has gone out, even though it continues to slowly rotate. They have crossed the death line and faded into the nothingness of space, joining the graveyard of dead neutron stars.[9]

Accretion-Powered Pulsars: Hungry Zombies

Not all rotation-powered pulsars stay dead. Some have a gravitationally bound companion star that they can cannibalize, and so will wake up from the grave; scientists call them recycled millisecond pulsars.

December 29, 2016, was an unusually snowy and cold day for the Netherlands. It was months before the supercomputer at the University of Groningen would capture data from the slowest-ever pulsar, but LOFAR still managed to surprise astronomers with a belated Christmas present. Cees Bassa, a researcher at ASTRON, decided to check LOFAR's latest observations—after all, the telescope wasn't on holiday and was continuing to search for pulsars. Sitting in his bedroom and casually sifting through the latest data, Bassa suddenly spotted a peak in the signal—possibly a pulsar unlike any other in the survey, spinning at incredibly high speeds. Thrilled, and disregarding the winter break, he shot an email to a colleague, Jason Hessels, at the University of Amsterdam. The subject line was a teaser: "Say hello to . . .". When Hessels opened the message and saw the data, the holiday was over for both of them.[10]

The new pulsar, now called PSR J0952 – 0607, is the second-fastest pulsar ever recorded, nearly rivaling the record-breaker that Hessels had spotted a decade earlier, in 2004 (more on this discovery in Chapter 5). Both pulsars spin in the millisecond range; think of Bassa's pulsar as a ball the size of Washington, DC, spinning 707 times every second. As he read Bassa's email, Hessels "knew right away that it was real, even before we did any confirmation," he says. "I was super excited; for the last fifteen years, this has been one of the main things I've been interested in. So even if you're on holiday and something that exciting happens, you immediately drop everything and start thinking about it, because it's kind of the most enjoyable part of the job." He was not the only one to celebrate—he shared the news with his wife and their daughter, Dimphy, then seven years old. "I told Dimphy that there was this new star, and it was very special, because it was spinning around so fast that it must be very, very dizzy," he says. Dimphy started to spin around too: "Like that, Daddy? Am I a neutron star?"[11]

Pulsars spinning this fast are called millisecond pulsars. US astrophysicist Donald C. Backer discovered the first one in 1982, after his graduate student Shri Kulkarni sampled the data they had taken with the Arecibo Telescope and discovered one spinning at an outlandishly fast rate. Called PSR B1937+21, it rotates twenty times faster than the Crab Pulsar—nearly 642 times every second—and managed to hold the speed record for more than two decades.

So far, astronomers have detected more than three hundred millisecond pulsars, most of them in the Milky Way's galactic disk. Their density is the highest in globular clusters, crowded groups of old stars with hundreds of thousands, and sometimes even millions, of residents, but even so, only about 5 percent of millisecond pulsars live there.

Unlike radio pulsars that spin a few times per second all by themselves, millisecond pulsars are typically found in pairs with another star, usually a white dwarf. That's not how they start their lives, though. Binary systems like these originally consist of two ordinary stars that happily orbit each other. At some point, the more massive star exhausts its nuclear fuel and goes supernova, leaving a neutron star in its place. Assuming the system survives the supernova and neither the neutron star nor its companion is kicked out, the two will continue orbiting their common center of mass, with the pulsar spinning down and emitting energy in radio waves (but also occasionally in x-rays and gamma rays), powered by rotation. During its lifetime, though, this pulsar may very well slow down to the point of going silent. It will die and we would not be able to spot it any longer with our telescopes.

Billions of years later, the pulsar's friend—the regular, low-mass star—will near the end of its own life. It will swell up and become a red giant. And that's when things get really interesting. The swelling will move the star closer to its companion, to the point that the red giant's surface will become heated by the pulsar's emission, making its matter evaporate off the surface. When that happens, the companion will begin to fling a lot of debris in the direction of its dead friend, forming a disk of material around the neutron star in a process called accretion. The accretion disk is like a donut of hot, swelling matter that goes down the bath plug, round and round, eventually descending onto the neutron star's surface. This siphoning off of matter onto the neutron star somehow reduces its magnetic field, although that's a process no one understands yet.

The accretion transfers angular momentum to the neutron star, making it rotate more rapidly—its companion effectively breathes new life into it. The infalling material interacts with the neutron star's magnetic field, which eventually has enough magnetic force to overcome gravity and trap the hot gas plasma along magnetic field lines, until the gas plasma begins to stream down onto the poles of the neutron star. The neutron star then forms a hot accretion spot—simply known as a hot spot—just around the magnetic poles, like a hump or a mountain on top of the poles. The hot spot starts to emit radiation in x-rays—which is the moment when we can detect the pulsar again, provided the hot spot is facing our x-ray telescopes. The system is now called a low mass x-ray binary (LMXB), because the companion transferring matter onto the neutron star was originally a low-mass, Sun-like star and because the pulsar emits x-rays.[12]

Around two hundred LMXBs have been found in the Milky Way so far, thirteen of them in globular clusters. They can be observed with space-based x-ray telescopes such as NASA's Chandra X-ray Observatory and XMM-Newton. But their discovery started with the launch of a rocket.

The year was 1949. An x-ray detector was hitching a ride on a converted V-2 rocket lifting off from the White Sands Missile Range in New Mexico. It pierced the Earth's atmosphere, aiming to record x-rays from the Sun. Astronomers had suspected there would be x-rays coming from our very own star. Such radiation, they knew, would be absorbed by our atmosphere, but they expected stars and other space bodies like our Sun—containing extremely hot gases at temperatures from about a million Kelvin to hundreds of millions of Kelvin—to also radiate x-rays. The Sun's x-ray output was thought to be much smaller than the visible light emitted by our star; that's why astronomers thought they would never be able to detect any x-rays from more distant stars.

In 1962, they realized that their assumptions were wrong. Riccardo Giacconi, an Italian astrophysicist, had put another x-ray detector on yet another rocket, an Aerobee 150. It took off on June 12, 1962, and—for the first time ever—was able to observe cosmic x-rays that were clearly coming from a source, now known as Scorpius X-1 (Sco X-1), outside our Solar System. While this object was much farther away than our Sun, it was emitting a hundred thousand times more than the total emission of the Sun in all wavelengths. Sco X-1 was clearly not a star. Giacconi's detector also showed that the whole Universe was flooded by x-rays. The age of x-ray astronomy had begun.

Just as Jocelyn Bell in the mid-1960s was patiently hammering poles into the mud of Cambridgeshire, the Soviet astronomer Iosif Shklovsky was analyzing x-ray and optical data from Sco X-1. He suggested that the source of the radiation could be a neutron star accreting matter from a companion. Neutron stars were still only theoretical at that time, but not much later Bell found LGM-1 and its three cousins. Soon, Sco X-1 was confirmed to be an x-ray binary—specifically, an LMXB. It is still, to this day, the brightest x-ray source known in the sky.

In 1970, Giacconi's team launched Uhuru—the first x-ray satellite, which later proved the existence of black holes. Uhuru was actually the probe's nickname; officially it was known as the Small Astronomy Satellite-1 (SAS-1) and the "X-Ray Explorer." But Uhuru stuck: the word uhuru means "freedom" in Swahili, and the satellite was launched from San Marco off the coast of Kenya on the country's Independence Day. Later, Giacconi worked on the Einstein X-ray Observatory, the first imaging x-ray telescope launched in 1978, and then its successor, Chandra, launched in 1999.[13] "The technology wasn't fantastic back then, so you didn't get a proper x-ray image like you do these days from the Chandra or XMM space X-ray satellites," says David Buckley, an x-ray astronomer at the South African Astronomical Observatory in Cape Town. "In those days, you just had a sort of area where you knew the x-ray source was, so it was a real effort to try and locate the optical counterparts. People even used old-fashioned photographic plates to look for blue objects and variable objects in the sky."[14]

One of the first x-ray sources that was identified as a binary was Cygnus X1. Its name followed the traditional pattern: when astronomers first started discovering x-ray sources, they named them after the constellation where it was found, and added an "x" to show it was an x-ray source. Eventually, researchers gave up on that naming system, because they have now found millions of x-ray sources. Today, modern detectors like INTEGRAL, Swift, NICER, and Maxie (a small experiment run by Japanese scientists on the ISS, the International Space Station) are all looking for sudden eruptions of x-rays in the sky.

Apart from LMXBs, there are other types of neutron stars that emit in x-rays. If the companion star in a binary is a medium-size star, the system is called an intermediate mass x-ray binary (IMXB). If the star is greater than ten times the mass of the Sun, you've got a high mass x-ray binary (HMXB). In the case of the HMXB, one star will go supernova and become a neutron star. The companion, though, is an extremely luminous star that emits stellar wind because of radiation pressure, with the neutron star siphoning the material not to an accretion disk but directly onto the surface. The energy of the wind gets converted into x-rays. It is also possible to see HMXBs in the optical light, because their emission is dominated by the massive companion star. Not all HMXBs contain a neutron star, though; sometimes the partner is a black hole.

But it's LMXBs that are arguably the most unusual—because these systems are thought to be the progenitors of ultra-fast and very, very old millisecond pulsars.[15]

As accretion progresses, with the material from the companion still raining onto the neutron star, the star's magnetic field keeps weakening. When it gets as low as about 10^8 gauss, the accreting debris is so close to the surface that it makes the pulsar spin at millisecond speeds. Once accretion ends, x-ray emission ceases—and the pulsar is now a millisecond pulsar, emitting in radio once again, in a process called "recycling." This combination of a weakening magnetic field and an accelerating spin causes the pulsar's lifetime to increase: rather than lasting for ten to a hundred million years like a visible pulsar, this neutron star will now live for a substantial fraction of the age of the Universe, more than a billion years. Say hello to the twice dead, "zombie squared" leftover core of a once massive star that has morphed into a very old, rotation-powered pulsar.

Its companion, meanwhile, has turned into a white dwarf, and will either stay a white dwarf or can get ablated completely by strong "winds" of high-energy particles from the pulsar, which happen when the pulsar heats the companion to more than twice the temperature of the Sun's surface, eventually eroding it. This is why some millisecond pulsars are not in binaries but single—like the very first millisecond pulsar found by Backer. They are called "black widow" pulsars—just like the spiders that consume their mates. So far, we have found eighteen in the Milky Way and a few more in globular clusters that orbit our galaxy. Some of them don't have companions, while others have a very, very low mass companion star. That's exactly the type of system that Bassa spotted—with a white dwarf companion for his millisecond pulsar that has merely 2 percent of the mass of the Sun. It clearly has lost much of its matter to its neighboring and very hungry pulsar. When the companion is slightly more massive, but still clearly struggling, the pulsar is called a redback—another spider analogy.[16]

While millisecond pulsars typically emit in radio, some can't make up their mind and periodically switch between x-rays and radio emission. These are called transitional millisecond pulsars, and they are peculiar beasts. In 2008, a team led by University of Amsterdam astronomer Anne Archibald discovered a radio pulsar now called PSR J1023+0038, using the Green Bank Telescope in West Virginia. When she and her colleagues checked archival data, they realized that about eight years earlier in this exact spot a neutron star had been visible in optical wavelengths, showing an accretion disk. They started monitoring the new pulsar with the Lovell Telescope, Arecibo, Green Bank Telescope, and the Westerbork Telescope; the pulsar could be observed until June 2013, when it suddenly vanished. Weeks later, the accretion disk was back and the star was visible in optical wavelength again, incredibly bright because of the accretion disk around it. Later observations in other parts of the spectrum, such as with x-ray observatories in space and optical telescopes on the ground, showed that the system is changing between, on the one hand, broadcasting in radio, and on the other, emitting in x-rays while accreting matter and being visible to the human eye.[17]

Very rarely, such binary systems are made up of two neutron stars and both are pulsating. So far, we know of only one such system—the Double Pulsar (PSR J0737 – 3039), although pulses from the slower one, called Pulsar B, haven't been detectable since 2008. Its partner, Pulsar A, is still happily emitting in radio—spinning at millisecond speeds.[18]

Magnetars: Strongest Magnets in the Universe

Another, even more powerful type of emission mechanism involves the magnetism of some neutron stars, which is so extremely strong that these neutron stars are probably the most highly magnetized objects in the entire Universe. As their magnetic fields decay, these neutron stars emit radiation in x-rays and gamma rays, which we can observe. They are called magnetars—and so far, scientists have discovered only about thirty of them.

"Magnetars were discovered serendipitously," says Chryssa Kouveliotou.[19] Now an astrophysics professor at George Washington University in Washington, DC, her work on magnetars started back in 1979, even though she didn't realize it back then. As a graduate student from Greece working at the Max Planck Institute in Munich, Germany, she was fascinated by the mystery of the phenomenon called gamma-ray bursts (GRBs).

Kouveliotou decided to write her PhD thesis on these ultra-energetic flashes coming from deep space, which had been first observed back in July 1967. Two US military reconnaissance satellites called Vela, built to spot gamma radiation from nuclear bomb explosions and orbiting Earth to ensure that nobody violates the treaty banning nuclear weapons tests in space, suddenly detected a brief flickering of gamma rays, which each lasted just two tenths of a second. The signals were strikingly different from what a weapon made on Earth could possibly have produced. In three minutes, the flashes were gone. Then, fourteen and a half hours later, another, weaker beam of x-rays arrived from the same location—high up in space. Thanks to the Earth's atmosphere, no one—apart from the scientists working for the US government space missions—noticed the bursts. Where had they come from? The US military was alarmed.

Later, other detectors recorded similar bursts. In total, sixteen were spotted and studied, first as a top-secret matter at the Los Alamos National Laboratory, a US Department of Energy lab set up during World War II to develop nuclear weapons. A team of Los Alamos researchers finally published their discovery in 1973, stating that the signals were clearly extragalactic and had nothing to do with weapons on Earth. Still, nobody knew where they might have originated. Kouveliotou in her thesis argued for what was then the most popular theory—that the flashes had been produced during the collapse of an ultra-massive, rapidly spinning star into a black hole, during the final stages of its evolution. This theory is still in line with current thinking when it comes to long gamma-ray bursts (longer than two seconds). Much briefer flashes, however, so-called short gamma-ray bursts, were shown in 2017 to be triggered by the merger of two neutron stars.

Because of their extreme brightness, many researchers first assumed that gamma-ray bursts had to originate from not too far away, somewhere within our own galaxy. That would bring the intensity of the observed gamma rays and x-rays just below their Eddington limit (the maximum luminosity that a very hot and bright star can achieve given the competing forces of outward radiation and inward gravity).

And then something surprising happened. On March 5, 1979, a powerful wave of gamma radiation washed over Venera 11 and 12, a pair of Soviet spacecraft orbiting the Sun. A few months earlier, both spacecraft had dropped probes into the acidic toxic atmosphere of our next-door planetary neighbor, Venus. Russian astronomers monitoring the spacecraft's sensors for galactic radiation had just recorded a typical reading: around a hundred counts per second. But when the wave of gamma rays hit in the early evening Moscow time, radiation levels soared above two hundred thousand counts per second, saturating the instruments.[20]

The wave didn't stop there. Eleven seconds later it reached NASA's Helios 2 craft, also orbiting the Sun, then washed over Venus, where it flooded the detector of a US probe, the Pioneer Venus Orbiter. Earth was next. The beam zapped the detectors on three US Department of Defense Vela satellites, the X-ray Einstein Observatory—the first fully imaging x-ray telescope—and the Soviet Prognoz 7 satellite. Finally, on the way out of the Solar System, it reached the International Sun-Earth Explorer 1 (ISEE-3), a probe sent to study magnetic fields near Earth. Luckily, a year earlier, shortly before the launch of ISEE-3, several scientists studying GRBs had asked to piggyback a couple of gamma radiation detectors on the craft: the field was very new and there were no designated gamma-ray probes yet, but their request was granted.[21]

Kouveliotou was at her desk at the Max Planck Institute for Extraterrestrial Physics when the data from ISEE-3 started rolling in. She still had a year or so to go before defending her thesis. At first, she recalls, everyone thought that the peak was the result of an instrumental fault—but the detection was quickly confirmed as the data from multiple satellites came in.

The gamma-ray wave washing over our Solar System turned out to be about a hundred times stronger than the gamma radiation that had hit the Vela satellites back in 1967. The burst was just a fraction of a second long, but it turned out to carry as much energy as the Sun emits in ten thousand years. Kouveliotou and her colleagues were stunned. All GRBs detected until then had seemed to be the result of catastrophic, one-off events, with the source vanishing in the explosion. But now, on March 5, 1979, this first peak was followed by a hundred-second tail, with periodic, repeating peaks. The periodicity of the tail was especially startling. "That meant that the compact object had a surface—so basically, that was a neutron star," says Kouveliotou. Soon after, more bursts were detected, albeit less intense. But the signals were clearly coming from the same direction in the sky and didn't agree with any of the theories about the origin of GRBs. "That was the puzzle, and we were at a loss," recalls Kouveliotou. Some scientists even suggested that the first flash might have been triggered by a comet crashing into a neutron star.[22]

Not long after, the origin of the March 5 event was traced to the Large Magellanic Cloud—and associated with a young, roughly five-thousand-year-old supernova remnant called N49. This meant that the source was about a thousand times farther away than originally assumed based on its brightness. Its luminosity also had to be at least a million times over the Eddington limit. It couldn't be coming from a black hole, because of the periodicity of the signal's "tail." And then there was the supernova remnant. Was the source even a neutron star? At the time, pulsars were known only as quickly spinning neutron stars emitting radio waves, but the x-ray burst observed on Earth was too bright for what could be expected to come from a rotation-powered pulsar. The neutron star, if that's what it was, was also not in the center of the remnant, but on its outskirts, which indicated that at birth it had been ejected from the progenitor star's location at a speed of some one thousand kilometers (about 600 miles) a second—much faster than any other known pulsar at the time.

Over the next four years, Russian astronomers at the Ioffe Institute in Saint Petersburg detected sixteen more bursts coming from the same direction—some stronger, some weaker, but all fainter and shorter than the March 5 flash. Later, several other similar events were registered, coming in bunches from three different directions in the sky. Everybody was excited, but no one knew for sure what they were.

Kouveliotou graduated in 1981, and later, after a stint at the University of Athens as a lecturer, landed a job at NASA's Marshall Space Flight Center. GRBs and the mystery burst from 1979, however, stayed on her mind. When in 1986, at an astronomy conference in Toulouse, researchers started discussing the 1979 events, she had to speak up. Most scientists at the time assumed that the events were a type of GRB, but Kouveliotou

Chryssa Kouveliotou
(Courtesy of Chryssa Kouveliotou)

did not believe they were. She had done her thesis on GRBs and was certain this wasn't one.

Opinion was still very much divided, but at that conference it was decided to call them soft gamma repeaters (SGRs), because their energy was not as high as those of typical GRBs, and, since they were repeating, they were clearly not associated with any catastrophic phenomenon. More and more scientists also began to agree that while GRBs were much more powerful and originated deep in space beyond our own galaxy, the SGRs seemed to live much closer—along the plane of the Milky Way.

The source first observed on March 5, 1979, continued to burp out energy occasionally, with a perhaps final burst spotted in May 1983. Kouveliotou set out to solve this mystery. But she had a problem: there still weren't any tools to probe the nature of the mysterious bursts, and there was no high-resolution imaging available yet to locate them. For more than a decade, there was not much Kouveliotou could do but wait. Little did she know that a couple of theoretical physicists had also begun to approach this mystery from another direction. Soon their paths would cross.[23]

CHRIS THOMPSON WAS STILL IN HIGH SCHOOL when that wave of gamma radiation hit Earth in 1979. Unlike Kouveliotou, he had no idea then that GRBs even existed. But by 1986, still a graduate student, he and postdoc Rob Duncan—both at Princeton University—became fascinated by the magnetic fields observed in radio pulsars. They wondered how magnetic fields made pulsars spin down, and why some pulsars are more magnetized than others.

An important clue came from a study of newly formed neutron stars that had recently been published by Adam Burrows of the University of Arizona and James Lattimer of the State University of New York at Stony Brook. They developed computer simulations suggesting that the dense fluid inside a still-hot neutron star circulates by convection for a few seconds as it begins to cool. All stars have weak magnetic fields deep inside them and transfer some residual magnetic field to the neutron star. But, Duncan and Thompson reasoned, sometimes these fields may get much, much stronger. This wouldn't happen with every pulsar, but only with those that were rotating extremely fast at their birth—triggering the so-called dynamo effect. This effect operates in our own Earth and in most stars. It happens when an electrically conducting fluid or gas swirls around, with hot parts rising to the top then cooling off and sinking again, just like boiling water in a pot. The magnetic field is effectively tied to charged particles in the fluid, so it is stretched and amplified by the fluid motion.

At this stage the inside of a neutron star is still an ordinary fluid, not yet a superfluid, which forms only when the star cools significantly. This hot neutron fluid, at some 30 billion kelvins, will bob up and down, at speeds of thousands of kilometers per second. If, they argued, the initial magnetic field of a newborn neutron star is strong enough and the spin is fast enough—more than two hundred rotations per second—then eventually a dynamo effect would be triggered. Although lasting only a few seconds, that would be enough to amplify the magnetic field to more than 10^{15} gauss: a thousand times stronger than the magnetic field of a typical neutron star, which rotates too slowly for the convection to start and the dynamo effect to kick in. Recall that the Earth's magnetic field is roughly a mere half a gauss, a fridge magnet is about a hundred gauss, and sunspots—the most magnetic areas of the Sun—are some three thousand gauss. The stronger the initial magnetic field, the faster such a neutron star will die, eventually rotating far too slowly to emit radio waves; this extinction of a pulsar usually takes between 10 million and 100 million years.

Thompson and Duncan decided to explore further the amplification of the magnetic field. "We got thinking about how these fields would manifest themselves—whether there were any observational signatures of them," says Thompson. In 1992, the duo published a bombshell paper that coined the term "magnetars," or magnetic stars, to describe such peculiar objects: neutron stars with magnetic fields stronger than any other known object in the Universe. They even calculated an upper limit to the magnetic field that a neutron star could possibly produce and maintain, of about 10^{17} gauss; go over it, and the nuclear fluid inside would start to mix and the field would weaken.

The theorists also explored the effect of an ultra-strong magnetic field on a neutron star's rotation. They quickly deduced that such a field would slow its rotation dramatically and rapidly, with the magnetic energy eventually dominating whatever energy was left over in the rotation of the star. While such a magnetar would be born spinning faster than a typical pulsar, its spin rate would plummet much more quickly, with the magnetic waves carrying off the star's rotational energy—reaching a period of eight seconds after five thousand years or so for a field of 10^{15} gauss. Furthermore, magnetars don't spin down at a constant rate; rather, they spin down very rapidly at first, doubling their period in minutes to hours, and the rate of spinning down gradually decreases. Neutron stars are renowned for being stable clocks, especially the millisecond pulsars—but magnetars are terrible clocks; the rate at which they spin down wanders all over the place, surging and fading even by a factor of ten.

As it happened, eight seconds was the periodicity in the event observed on March 5, 1979. This suggested a connection between soft gamma repeaters and the ultra-strong magnetic fields that Thompson and Duncan were finding in their dynamo theory. The 1979 burst turned out to be especially helpful for other reasons—because of its extreme energy and brightness. The theorists realized that an ultra-strong magnetic field could explain both the very intense brightness (at its peak, a giant magnetar flare like the 1979 event is a thousand to ten thousand times brighter than the short, repeating SGR bursts), and the long tail of regular pulses that had been detected following the bright gamma-ray peak. Because the magnetic fields and the remainder of the outburst were dragged along as the neutron star rotated, every time it faced the Earth, astronomers detected a pulse.

The two physicists argued that it was possible to calculate the strength of the magnetic field for this neutron star in two different ways. The first, as described, combined the observed spin period with the age of the supernova remnant N49. But the huge energy of the 1979 burst, if powered by a magnetic field, also required that the field be stronger than 10^{14} gauss. Working backward, they reasoned that when such an extremely strong field drifts through the solid, typically stable crust of a neutron star, twisting and turning—it would eventually apply so much stress that the crust would crack, causing a "starquake." This would twist the magnetic field outside the star, energizing clouds of electrons and positrons, and triggering a tremendous outburst of magnetic energy in the form of hard (high-energy) gamma rays—similar to but much more powerful than a solar flare coming from inside the Sun.

The pulsating tail, then, was powered by the residue of the dispersing and shrinking fireball, a hot cloud of electron-positron pairs that was trapped by the magnetic field close to the spinning neutron star. This residue would cool down as x-rays escaped from its shrinking surface.

"There's probably a process called magnetic reconnection involved in the energy deposition outside the neutron star," says Thompson. Magnetic reconnection happens when the Sun experiences an explosive event called a solar flare and magnetic field lines near the surface change how they connect to each other. This process releases magnetic energy—and solar flares often emit some energy in gamma rays. "But the details of how that happens in magnetars are very much a subject of active research," Thompson adds.[24]

In magnetic fields of such strength, all sorts of strange new phenomena emerge. X-ray photons merge and split in two, while atoms are thought to totally deform—becoming long and thin like spaghetti. Around the neutron star, the very hardest (highest energy) x-rays wouldn't be able to pass through as usual; its highly magnetized vacuum would act like a polarizing filter—letting only one direction of polarization pass through, just like sunglasses do. In 2016, this kind of polarization was confirmed observationally. Thompson and Duncan also theorized that how rapidly a magnetar was spinning down depended on the star's magnetic activity. In 1995, the two scientists published another paper expanding on their theory, outlining seven different ways to calculate the magnetic field of the March 5 burst. All seven approaches yielded the same result: a magnetic field stronger than 10^{14} gauss.[25]

Many astronomers were skeptical about Duncan and Thompson's bold claims, but Kouveliotou was not among them. All these years, she had been waiting for better instruments to go into orbit. Was the 1979 burst really a SGR? And what was a SGR, anyway? When she read the paper suggesting the existence of magnetars, she and a small team became determined to test the concept.

On December 30, 1995, NASA was getting ready to send a rocket to space, carrying the Rossi X-ray Timing Explorer (RXTE) satellite. The probe had a very specific aim: to monitor how various cosmic x-ray sources changed as time went by, focusing on x-rays from black holes and neutron stars. Kouveliotou wrote a proposal to observe the soft gamma repeaters that she hoped would be activated during the RXTE operating time. Her idea was accepted. And as luck would have it, shortly after the satellite arrived at its designated orbit around Earth, its instruments did indeed register a pulse from one of the sources that had burped in 1979.

Two years earlier, in 1993, the Japanese ASCA satellite had already observed this SGR and determined its precise location but had not identified a period for the star; now Kouveliotou and her collaborators were able to show that its x-ray emissions were pulsating on a regular 7.5-second cycle. As the neutron star was spinning, hot spots in its magnetosphere were rotating in and out of view—so 7.5 seconds was the period of the star, very close to the eight-second periodicity of the first burst detected on March 5, 1979.

Kouveliotou started monitoring the object, aiming to figure out the period and the spin-down rate to then calculate the strength of the body's magnetic field—and to see whether Duncan and Thompson were right in their predictions that an extremely strong magnetic field would make a magnetar slow down very fast. She effectively started "timing" it. Timing is a widely used technique for observing pulsars; researchers carefully monitor the arrival times of pulses from one or several pulsars over a few years. They hope to spot any changes in the arrival time of the pulses, because these could be signs of some disturbance in the fabric of space, for example, when a neutron star and its binary companion orbit their common center of mass. (For more on this technique, see "Deeper Dive: Pulsar Timing.")

But because the source data were noisy and fainter, Kouveliotou's timing technique was slightly different from how radio astronomers routinely time pulsars. Radio pulsar measurements are incredibly precise, and it's possible to connect up the phase of the rotation at very different times. For instance, suppose that the radio brightness shows two peaks every rotation, a big peak and a small peak. Then it's possible to measure the spin period so precisely that one can very accurately project into the future when the big peak will arrive. When a second measurement is made days or months later, the arrival time will be slightly shifted—because the star is spinning down. But the method Kouveliotou used was somewhat simpler—based on a technique that x-ray astronomers use when timing accreting x-ray pulsars. The periodicity in the x-ray signal due to the rotation is measured within a window of a few hours. Then, at the later date, the same measurement is made again. But because the precision of that measurement is lower than for radio pulsars, it is harder to connect the phase of the rotation between the two epochs.

Sure enough, over half a decade of timing, the SGR slowed by about two parts in a thousand, or one second every three hundred years, which was faster than that of any known radio pulsar. She did the calculation and found that if the star's slowing down was caused by magnetic fields siphoning away energy and angular momentum, then its field strength had to be 8×10^{14} gauss. This was very close to what Duncan and Thompson had estimated and higher than any known magnetic field of any pulsar or any other object in our Universe that we know of. "The precision of Chryssa's measurement was good enough to show to high significance that the SGR was spinning down," says Thompson. "That was enough!"[26]

When Kouveliotou got the data, she was thrilled—but to make sure it was watertight, she sent it to two other groups to blind check, without telling them the period that she had determined. They, too, calculated the same period and its rate of change. Now the importance of her discovery dawned on her. Sitting in NASA's Marshall Space Flight center in Huntsville, Alabama, she contacted Duncan and Thompson. "We were now completely sure that this was real, and we were just exhilarated," she says. "We started writing the results down for *Nature,* and that's the story of it. And we started calling them magnetars, because they were the first example of a magnetar."[27]

The *Nature* paper came out on May 21, 1998, and stirred a lot of interest in the astrophysics community. "For me, this is something really monumental. It was a tremendous experience to be the first one to find a source," she says. "I always kept saying that this was something new. And I kept saying this until I finally measured their magnetic fields and it did come up as something new because it had a very high magnetic field."[28]

That was just the beginning. A few weeks later, SGR 1900+14 burped more than fifty times, and in June 1998, astronomers spotted a fourth source, SGR 1627−41, which produced powerful flashes some one hundred times over the following two months. On the morning of August 27, 1998, an ultra-powerful flare of gamma rays and x-rays washed over the Earth. It was even stronger than the one on March 5, 1979, and flooded the detectors on seven satellites spread throughout the Solar System; it even triggered a sudden change in the Earth's upper atmosphere called the ionosphere, impacting radio communications. At 3:22 a.m. PDT, the inner border of the ionosphere suddenly plunged, for five minutes, from the usual nighttime altitude of 85 km (53 miles) to 60 km (37 miles), where the ionosphere is usually found during the day, when high-energy photons from the Sun keep the air at this level more ionized. This sudden drop was caused by the second giant flare of gamma rays detected from the magnetar that burst in 1979, SGR 1900+14. Importantly, Kouveliotou's group also detected rapid spindown in this SGR, implying a similar magnetic field to that inferred in the first spindown measurement. For Thompson, the confirmation through observation felt especially crucial. "As theorists, I think we were very happy that our work was of some use in planning observational campaigns. Theorists in astronomy usually are the sweepers-up; we try to systematize or explain things after the fact. It's less often that our work turns out to be important in suggesting new observations that otherwise wouldn't be carried out."[29]

While the scientific search to explain soft gamma-ray repeaters had been going on, x-ray astronomers had been perplexed by another phenomenon. A year before the March 5, 1979, flare hit Earth, NASA had launched the Einstein X-ray Observatory, the first satellite to image x-rays from distant objects. The probe, just like many others, was flooded by the wave from the SGR—but survived without permanent damage.

That same year, however, the Einstein X-ray Observatory found something else that was new. Astronomers had already known about a number of x-ray pulsars, and they knew that x-ray pulsars all seemed to exist in pairs, emitting radiation via accretion of matter from their companion.

In December 1979, Philip Gregory and Greg Fahlman of the University of British Columbia in Canada used Einstein to look at some radio pulsars—and spotted a bright point x-ray source in the constellation of Cassiopeia, some ten thousand light-years from Earth. It was surrounded by what seemed to be a gas cloud, shining brightly in x-rays. They called it 1E 2259 + 586 and theorized that the object was a neutron star in its supernova remnant home, shining in x-rays with a luminosity of a hundred Suns.

Despite a lot of searches, astronomers never found a companion for the Cassiopeia star, so accretion could not be its power source. Neither could its emission be coming from its rotation, because there were no radio waves, and while it was spinning once every seven seconds, the rate at which it was losing energy and slowing down was much too small to explain the extreme x-ray brightness. The new pulsar's emission mechanism was a mystery. Astronomers kept searching for a companion with ever-better technology, but in vain—there was no evidence of it either from deep optical observations or from attempts to use x-ray timing to look for an effect known as Doppler shift due to binary motion.

Instead, astronomers spotted three more solitary, very bright x-ray emitters, each with pulse periods close to six seconds. (In contrast, strongly magnetized neutron stars that were accreting in x-ray binaries were spinning much faster.) With four new, strange beasts to consider, it was time to declare a new class of neutron stars. At a March 1995 conference in La Jolla, California, Thompson and Duncan suggested that these pulsars could also be magnetars—and dubbed them Anomalous X-ray Pulsars (AXPs).

The two astronomers elaborated that young magnetars would first emit in radio, rotating and spinning down fast—a sort of a transitory flywheel. The electric currents that support their intense magnetic fields would then slowly decay over thousands of years, causing them to emit brightly and continuously in x-rays. For years, this was just one of several theories. Duncan and Thompson reasoned, though, that if AXPs were indeed also magnetars, just like SGRs, they should occasionally have bright outbursts that astronomers should be able to detect. To observe such a phenomenon, though, researchers really needed some new technique—and Vicky Kaspi, an astronomer at McGill University in Montreal, had an idea.

Waiting for the "Glitch"

I meet Kaspi on a rainy October day, the maple trees outside her McGill office competing with each other to display the most vibrant fall shades of yellow and red. She is a slim, petite woman with beautiful curly hair and a seemingly permanent smile. Kaspi has a wealth of experience observing radio pulsars—as part of her graduate studies, she did a lot of pulsar timing, particularly millisecond pulsars using Arecibo Observatory. Those were the very early days of monitoring a so-called pulsar timing array—recording very precisely the arrival of pulses from multiple pulsars scattered across the sky. She knew exactly how to "keep phase"—to keep track of every rotation of a neutron star, over long periods of time. "I knew that the key was to have two closely spaced observations, within an hour or two, followed by a third a few hours later—and then at increasing spacings until you could observe them just monthly," she says.[30]

But that was radio. One day in 1995, still a postdoc at Caltech, she happened to hear physicist Tom Prince give a talk about AXPs. She found these weird accreting binaries fascinating. Duncan and Thompson had just suggested that they could be magnetars, and Kaspi wanted to help solve the mystery. The RXTE satellite had just been launched a few months earlier and she thought it would be interesting to see if she could apply the same phase-keeping technique she had used with radio pulsars to an AXP. She didn't know much about RXTE, but a friend and colleague at MIT, Deepto Chakrabarty, did—so they wrote a proposal together. "It was very demanding of telescope resources, but they approved it," she says.[31] And her timing method worked, as was obvious from the very first batch of RXTE data. Her technique showed that AXPs were fairly stable rotators, unlike accreting pulsars, which are quite unstable.

For Kaspi, it was proof that AXPs were not accreting from a binary companion. Still, it was not yet proof that they were magnetars. So she kept monitoring them. The timing technique, which was all about maintaining phase coherence, enabled her and Chakrabarty to detect sudden tiny changes in the rotational periods of these objects, and one day, the period changed. This was the first observation of a so-called glitch of an AXP, when the period of a pulsar—usually a radio pulsar, an extremely accurate celestial clock—suddenly becomes shorter, meaning the neutron star starts rotating faster, then returns to its original spin (for more on glitches, see Chapter 5). "This was important because until then only radio pulsars showed glitches, and this was further strong evidence that they had more in common with isolated radio pulsars than with accreting pulsars," Kaspi says.[32] These objects also had much higher magnetic fields than stand-alone radio pulsars, as was clear from the spin-down rates that Kaspi and Chakrabarty had measured.

The two researchers kept monitoring five AXPs on average every two to three weeks, leading to a fairly large data set. Another AXP glitched, but generally all were quite stable—and "to me this was enough proof, but not to everyone," Kaspi says.[33] At a conference in Boston in 2001, a colleague, Pete Woods, suggested to her that they should comb their data for soft-gamma-repeater-like bursts. So Kaspi, by then a professor at McGill, asked one of her PhD students, Fotis Gavriil, to do just that. The tip paid off—in September 2002, Gavriil showed her a likely signal. It was a bright x-ray flare coming from the direction of a known AXP called 1E 1048.1−5937. After a lot of analysis and discussion, the team concluded that it was most likely an SGR-like burst. It was not as bright as soft gamma repeaters, but brighter than regular x-ray pulsars. The team published the results in *Nature*.[34] Still, Kaspi worried that their analysis might be wrong—after all, they had only one source to go by, even though it had flared twice, with a sixteen-day break between flashes.

As it turned out, Kaspi's team had been spot-on. A few months later, in June 22, 2002, her cell phone rang. It was Jean Swank, the principal investigator of RXTE, and she sounded excited. "One of your AXPs is bursting and setting off alarms on XTE. What do you want us to do?" she blurted out. The source, 1E 2259+586, was in a completely different direction from the burst she had earlier reported in her *Nature* paper. Immediately, Kaspi replied: "Keep observing it!!" They did—and that's how they caught the first major AXP outburst, much brighter than the previous one. The source emitted more than eighty short bursts in total, and simultaneously glitched. That was not the behavior of a typical x-ray pulsar. "The bursting was so amazingly SGR-like that that was all the proof the community needed that AXPs were also magnetars," she says. The researchers even wrote in the title of their paper that the detections were from the "No-Longer-so-Anomalous X-ray Pulsar 1E 2259 + 586."[35] In hindsight, she says, the researchers were bold for claiming that the first burst was magnetar-like, but with the second one it was a no-brainer.

And adds, after a pause: "By the way, I was nine months pregnant at the time of that outburst, when the word 'burst' had a very different meaning to me!"[36] Her daughter was born about a week later, with Kaspi coordinating observing runs to study the outburst on several telescopes up to the very last minute. One colleague even suggested that she should name the baby 1E 2259 + 586; Kaspi decided to call her Julia instead.

Vicky Kaspi had brought her knowledge of phase-coherent timing of radio pulsars to the x-ray world. The timing showed that AXPs are very stable rotators, but they glitch, just like radio pulsars, and experience powerful outbursts—like SGRs. Since then astronomers have observed many AXP outbursts, completely blurring the line between AXPs and SGRs to the point that Kaspi and many others now refer to both as magnetars.[37]

Thompson, however, prefers to maintain the two categories. AXPs are discovered primarily by searching for persistent x-ray sources, he says, and none of the known AXPs has ever produced a really bright outburst like SGRs do. They tend to generate intermediately bright bursts that are detectable simply because someone happens to be pointing an x-ray telescope at them when they go off; otherwise they are usually too dim to be detected by our wide-angle gamma-ray detectors, which are designed to spot really bright bursts.[38]

Currently, only a handful of AXPs are known. One, also in Cassiopeia, was even detected in visible light. It's very faint, fading in and out as the pulsar rotates.

Until 2004, all of the known magnetars, about a dozen of them, had been detected and observed in x-rays or gamma rays only. They all were rotating slowly and relatively nearby, with most found in our own galaxy. A number of radio astronomers became intrigued and decided to point radio telescopes at these magnetars—to see whether, perhaps, they would also emit radio waves. But they didn't. By about 2005, some physicists even started writing papers trying to explain why radio emission from magnetars couldn't occur. "People were saying that the magnetic field is too strong compared to ordinary pulsars, so there would be some quantum mechanical effects preventing radiation," says Fernando Camilo, astronomer and chief scientist at the South African Radio Astronomy Observatory (SARAO), which manages all radio astronomy facilities in South Africa.[39] I met him at SARAO in Cape Town.

Camilo was at Columbia University back then, and across from him in the office sat astrophysicist Jules Halpern. One day in 2005, Halpern and a few colleagues found a point source in radio, without pulsations, in the data they had been analyzing from a Galactic Plane survey done by the VLA in New Mexico. The point source of radio waves was at the same location as an x-ray magnetar called XTE J1810−197, which had been discovered by the XRTE in 2003. Halpern wrote a paper in which he argued that the point source couldn't be a radio pulsar; he suggested that there had to be some other phenomenon at work, say, the shock of the ejecta from this magnetar expanding in the circumstellar medium. Camilo bought the argument.

But having submitted the paper, Halpern couldn't stop thinking about it—the source kept bothering him. So just a few weeks later, he spoke to Camilo—who was a radio astronomer while he wasn't—and asked him to point the Parkes telescope at this location. Camilo wasn't too keen: he, after all, had been a part of the large Parkes pulsar survey of the Galactic Plane, together with Andrew Lyne in the late 1990s, which hadn't seen any pulsar in that location. Halpern, however, refused to give up and kept asking Camilo to follow up. Finally, in March 2016, in part to get Halpern off his back but also "because Jules was often right," laughs Camilo, he asked the Parkes telescope operations scientist John Sarkissian to type in the coordinates. Sarkissian did, and Parkes swiveled its giant dish toward the x-ray magnetar for ten hours. "I remember it was a weekend, I was analyzing the data. Jules was at home because he didn't work on the weekends in the department, but I did often. So I was looking at the data, and bam! It was the brightest pulsar I've ever detected in my life," says Camilo, sitting in his office.[40]

The neutron star had 5.5-second pulsations, corresponding to the rotation period of the magnetar. They were ridiculously bright, recalls Camilo—so bright that his code automatically flagged them as RFI, or radio frequency interference. But he knew perfectly well what he was looking at, once he realized that these peaks were separated by 5.5 seconds. He immediately phoned Halpern. "I was so excited that I called him at home. I said, Oh my God, you're not gonna believe what I found. And he said, Oh, you found the magnetar!"[41] It was the very first magnetar detected shining in radio—a very unexpected find for the pulsar community.

The two physicists later analyzed why the source hadn't been spotted in radio earlier. It turned out that before 2002, the neutron star was emitting x-rays but very weakly, and suddenly the emission became a thousand times brighter. Something similar had clearly happened in the radio, Camilo thinks. So the point source was not just a point source after all, but emitted pulsations from the magnetar. The astrophysics community was stunned—there had to be, then, other magnetars emitting in radio, even though clearly they were not doing it at the time.

The magnetar's radio emission, however, was different from that of ordinary radio pulsars. A typical pulsar has a relatively steady and organized magnetic field, similar to a dipole—like a common bar magnet with a north and a south magnetic end—with radio emissions coming from around the magnetic poles and along the curving magnetic field lines. When the observed radio frequency increases by a factor of ten, the brightness (called the "flux density") typically drops by a factor of about a hundred, following a decreasing, approximately quadratic, relationship—although it differs slightly for every pulsar. If you go up in frequency by a factor of a hundred (say, 1 GHz to 100 GHz), the flux would go down by 100^2—or 10,000. But with magnetars, no matter how much the frequency changes, the flux remains about the same. "We went up to a couple of gigahertz, then to forty gigahertz, then we went up to nearly one millimeter observing wavelengths, which is crazy—we've never done that with any pulsar," says Camilo. "And we were still detecting single pulses from this magnetar very clearly."[42] Again the spectrum was very, very flat—meaning that the radio flux was more or less the same at all frequencies. And most of the radio energy was emerging at high frequencies (100 GHz), not at low frequencies (less than 1 GHz) as in ordinary radio pulsars.

In an ordinary solitary pulsar, the radio emission is due to the rotation of the neutron star; in contrast, magnetars emit x-rays and gamma rays because of the decay of the magnetic field. Those that also shine in radio are thought to emit radio waves because of their rotation, although the exact mechanism is not clear. Still, there must be some link to the magnetar activity of the crust, the magnetic field changing, and the star having hiccups—which would manifest in the changes of the radio emission. For instance, the first magnetar did not emit detectable radio waves until 2003, but they started at around the time of an x-ray outburst. "Clearly, whatever happened at the surface and in the interior of the star that led to the magnetar outburst in x-rays is related to what also caused the radio emission," says Camilo. Why? "We don't really know in detail whether it's this outburst that led to the x-rays and it created a lot of charged particles that are there and streaming around the pre-existing magnetic field, or is that the magnetic field itself gets rearranged and crunched up and somehow the acceleration that exists for any charged particle just becomes higher?"[43]

Camilo kept observing the magnetar for the next two years, when he noticed that in radio it started getting weaker and weaker. "I wish that I had observed that magnetar in 2002, or 2003, and 2004, and 2005, because who knows what we would have found. Maybe it might even have been brighter and more interesting," he says. "We only found the pulsations in 2006, which was already about three to four years after that x-ray outburst had occurred." The magnetar was already getting fainter in x-rays throughout that time, and when Camilo realized it, he noticed that the average radio flux was also decreasing.[44]

It died in 2008. Camilo was devastated, but he didn't want to lose hope. What if it were to suddenly revive? He had to wait—and in the meantime, he was scouring the cosmos for other magnetars emitting radio waves. In June 2007 he found another one that is still shining in radio more than a decade later. Then two more were discovered by other teams. The third one was spotted in 2009 during a Parkes survey covering the Milky Way; astronomers call it PSR J1622–4950, and it's considered close to home at just thirty thousand light-years from Earth. First they thought that it was just a pulsar, but then they realized that it was a slow pulsar, with a four-second period that was spinning down, which meant it had to have a large magnetic field, like a magnetar. They deduced that a magnetar-like point x-ray source was associated with it, and over the following years, they observed it with the Chandra x-ray telescope and indeed saw it in x-rays. And then in 2014 it died, too . . . for three years.

Camilo kept monitoring the original radio magnetar for the next several years, checking in roughly once a month for a few minutes, but it stayed dead. On April 26, 2017, he was already in South Africa working on MeerKAT, then under construction and with just sixteen dishes up and working out of what would eventually be sixty-four. He was about to leave his office for the Cape Town airport to board a plane to Australia when an email popped up. A colleague at Parkes who was monitoring magnetars for him had sent him a summary of the observations. "He was saying: 'Oh, Fernando, the first magnetar is still dead. The second magnetar, it's doing its thing, it's still bright. And the third magnetar is also very bright.' And I was like, what, but it's dead!" At first Camilo thought there had been a mistake—but realized that if not, it meant that the magnetar had suddenly turned back on. In the rush to make it in time for his flight, he emailed his colleague back, asking him to follow the magnetar up with more observations. But the colleague said that the Parkes Telescope was soon going to shut down for a one month to replace parts in its electrical system.[45]

"On the plane, I kept thinking, oh, this is a bummer," Camilo says. "It's a Southern source and Parkes is the only telescope in the world that can do it." And suddenly it dawned on him that he was the science director of a brand-new telescope. It wasn't fully ready, but so what? "When I got to Australia, I emailed some of my colleagues back in South Africa." They started observing for a few minutes every day for a month, looking at it "very obsessively," not just in radio but also with four x-ray telescopes, XMM, Chandra, Swift, and NuSTAR. It was shining slightly differently now, because the magnetic fields had rearranged themselves during their long slumber. But, says Camilo, it was amazing to have it back.[46]

The following year, the very first magnetar that Camilo had found, the XTE J1810−197, woke up, too. It revived in December 2018 after nearly a decade-long nap. A UK-German team working with the Lovell telescope at Jodrell Bank spotted radio emissions coming from the star with a spin of 5.5 seconds and an ultra-strong magnetic field of 10^{14} gauss. The magnetar was first found in x-rays in 2003, with radio emissions detected in 2006—and it kept emitting in radio until late 2008, when it fell asleep. So far, these are the only two magnetic beasts that have been seen to fall asleep and wake up again.

Another magnetar, discovered in 2013 and called PSR J1745−2900, is notable for its very peculiar location: near the supermassive black hole in the Galactic Center of the Milky Way. Its discovery was a shock: astronomers didn't think that they could detect even pulsars at relatively low radio frequencies that close to the Galactic Center. Not for lack of trying: telescopes have been scouting that area for over a decade, but in vain. Mainly though, the aim has always been to find millisecond pulsars and then use them to test general relativity as incredibly precise clocks moving in fast orbits around the Galactic Center. (For more on this effort, see Chapter 8.) This magnetar is not one of those: it spins with a period of 3.76 seconds and sports a magnetic field of 10^{14} gauss.[47]

Rare as they are, magnetars are incredibly mysterious. Astronomers do know that they fade as x-ray sources over several thousand years, but it's not clear how the magnetic field decays after the x-rays disappear—and it's possible that a substantial magnetic field remains frozen in the star, hardly decaying. Their radio emission is already irregular and probably becomes even harder to detect. So unless they revive, it then becomes very difficult if not impossible to detect them. "Searching for sporadic radio emission is a promising way of finding dormant magnetars," says Thompson.[48] Because detectable radio emission carries very little energy compared to x-rays, it will be up to new, much more sensitive radio facilities such as the Square Kilometer Array to discover them. They are definitely out there. Astronomers think that one is born in a galaxy such as the Milky Way every thousand years, and with about 100,000,000,000 galaxies in the Universe that means that some 10^{20} dormant magnetars lurk out there in the cosmos, waiting one day to wake up—and be noticed by someone watching the sky.

DEEPER DIVE: The Multibeam

More than a thousand of all known radio pulsars were discovered in the largest survey of the Galactic Plane to date—the Parkes Multibeam Survey of the Galactic plane, which used the veteran Australian dish. When Parkes started scanning the plane in 1997, just over seven hundred pulsars had been detected; barely a year later, on November 5, 1998, it bagged the thousandth known pulsar, making headlines around the world. The survey continued until March 2000 and then ran again in December 2001. The project's astonishing success was due to a very special instrument installed at Parkes in 1996—a multibeam receiver placed at the focal region of the dish. The multibeam was also used for the Swinburne University intermediate latitude surveys, and various others. In total, these surveys found well over half the known pulsar population, including the famous and so far unique Double Pulsar, mentioned in Chapter 2—the only known binary system that consists of two radio pulsars. The receiver also found the very first fast radio burst (FRB) dubbed the Lorimer burst—part of the new population of extremely powerful and brief flashes in deep space whose origin is so far unknown. (See Chapter 9 for more on FRBs).

Originally, the multibeam was developed to find very faint galaxies identified by the characteristic 21 cm spectral line of hydrogen atoms. At this exact wavelength, corresponding to 1420 MHz and falling within the microwave region of the spectrum, the radiation emitted by hydrogen gas penetrates the dust clouds that are opaque to visible light. But astronomers quickly reasoned that they could also use the multibeam to search for pulsars—thanks to the telescope's greatly increased field of view.

When John Sarkissian, the resident telescope operator at Parkes, showed me the multibeam, my first thought was: "Whoa, what a barrel!" Indeed, about a meter across and a meter and a half high, it does look just like a metallic barrel holding thirteen small cylinders inside, arranged in a hexagonal grid. It's a cryostat, which means that its designed to be very cold: when in use, the temperature inside is 20 Kelvin, not much above absolute zero, which eliminates noise and boosts sensitivity. For comparison, in outer space it's 2.73 Kelvin (−270.42 Celsius). When I visited, I saw the multibeam stored away in a tiny building next to the telescope. The astronomers had swapped it out for a single receiver, because Parkes was busy tracking the Voyager 2 craft on its journey out of the Solar System. For pulsar observations later, they would put it back in, hanging it high above the center of the antenna.

One of the people behind the design of the multibeam is veteran pulsar astronomer Andrew Lyne. It's a sunny Friday afternoon in July of 2019 when I meet him at Jodrell Bank Observatory. Astronomers had gathered for a summer barbecue near the 76-meter Lovell telescope just an hour's drive away from Manchester. Students, postdocs, and staff are sitting on the tarmac with the enormous telescope towering above, bathing in the sun. Compared to Parkes, where everything—from the control tower to the dish itself—says 1960s (in a charming sort of way), Lovell looks much more modern. Our phones are in airplane mode—radio silence is the rule here, just like at Parkes, although it's not easy given the dish's proximity to Manchester. I follow Lyne into a nearby building that looks more like a bungalow, where a few astronomers have their offices. Lyne has accomplished a lot during his long career, but one of the things he's most proud of, he says, is the huge Parkes survey and his role in creating the multibeam receiver.[49]

"We built the low noise amplifiers and bits of equipment that went into the cryostat," he says, adding that the back end of the Parkes receiver was actually built in Jodrell Bank.[50] Shortly after a multibeam receiver was installed at Parkes, in 1999, it was also placed on the Lovell, albeit with four feed horns. Before the multibeam, radio telescopes typically had just one feed horn. When radio waves from faraway objects reach a parabolic dish, they are reflected to the focal point of the antenna and enter the feed horn. The single horn collects the signals, which are then amplified in the receiver and converted into electrical signals. With this setup, the telescope can point at just one small spot of the sky at a time, which makes surveying larger areas very time-consuming.

That's where a multibeam helps. More feed horns clustered together allows the telescope to simultaneously observe several adjacent regions of the sky, thus covering the entire sky much faster than a single beam. The waves from each feed horn get into separate receivers—and because each horn has two polarizations, there are twenty-six outputs in all. Parkes's multibeam effectively turns the telescope a much larger dish, for a fraction of the cost. "Andrew Lyne had the initiative to use the receiver to look for pulsars and built rack after rack of analogue filters to find them," notes Swinburne University astronomer Matthew Bailes. "Eventually we upgraded the filterbanks to be digital for the HTRU surveys."[51] It's these surveys that later found the majority of the first thirty FRBs, convincing the world that these enigmatic radio flashes were real.

Since 2004, Arecibo Telescope in Puerto Rico has one too, with seven feed horns, as does the new five-hundred-meter Aperture Spherical Telescope (FAST)—a gigantic non-steerable dish in China. The FAST receiver has nineteen beams and a more compact design than the original one at Parkes; it was built by CSIRO and is the largest cryogenic receiver to date.[52]

DEEPER DIVE: The Exotic World of X-ray Sources

There's an odd subclass of extremely luminous and accreting pairs of astronomical objects that emit x-rays. They were first spotted in 2014, when astronomers detected pulsations from what they had assumed was a black hole, in a system classified as an ULX, or ultra-luminous x-ray source. The ULX had been known since the 1980s, when astronomers first observed extremely bright x-ray point sources that emitted more radiation than one million Suns at all wavelengths. The pulsations in 2014 showed that at least some of them are neutron stars, and to date, we know about six of them.[53]

Lee Townsend, an astronomer at the University of Cape Town, first noticed pulsations from one of these weirdos when it had an x-ray outburst in 2016. It was a known high mass x-ray binary in the Small Magellanic Cloud, and over the years that he had been studying it, it had displayed typical x-ray binary behavior—accreting just the way astronomers were expecting it to, with the emission they expected it to produce. Suddenly, it had a huge burst, becoming a thousand times more luminous than Townsend had ever seen, which put it into the ULX range. "This was one of the first times we actually observed a normal x-ray binary transitioning into the ULX regime," he says. "And it's one of the only pieces of evidence we have that x-ray binaries might actually be linked to ULXs."[54]

Townsend was taken aback, he says, by the power of the accretion—it was accreting at ten to twenty times the Eddington limit, the maximum luminosity that scientists thought a stellar body could achieve based on its mass. Even today, why it happened is a mystery. "Nobody really knows why these things are accreting so much," says Townsend.

DEEPER DIVE: Pulsar Timing

Typically, pulses of radiation produced by a standalone pulsar are so regular and precisely timed that on decadal timescales the best pulsars can rival atomic clocks for stability (although not on timescales of a day). Individual pulses occur at regular intervals and sum to a remarkably stable average pulse profile, a fingerprint that can be used to determine when the pulse struck the Earth.

But in some 10 percent of cases when a pulsar is in a binary system, moving with a companion, there are tiny but regular variations in when their pulses arrive at our telescopes. Timing is a technique for tracking the arrival of these pulses.

As the pulsar moves away from us, going around its companion and drifting farther from Earth, the pulses become redshifted, and fewer of them are detected every second. But when it comes closer to Earth, the pulses are compressed and more are detected per second. It's similar to the perhaps best-known example of the Doppler effect in our everyday world—when an ambulance is speeding toward us, the sound waves from the siren get compressed, meaning the wavelength gets shorter and their frequency (pitch) higher, with pulses closer together (weeoo-weeoo-weeoo). But the moment it passes us and starts moving away, the waves begin to stretch and we hear the siren in a lower pitch (weeoo . . . weeoo . . . weeoo . . .).[55]

Why do these variations in pulse arrival times occur? Radiation usually travels in a straight line—but according to Einstein's theory of general relativity, in the presence of a strong gravitational field, such as near a massive body that makes the fabric of space bend around its curve, radiation will follow the curvature and the photons' trajectories will also bend. The denser the object, the more it will bend the light near it. In 1919, Sir Arthur Eddington, director of the Cambridge Observatory, and Astronomer Royal Sir Frank Watson Dyson put the relativity theory to the test. With telescopes, they took an image of the patch of the sky centered on the Sun during a total solar eclipse, noting the specific positions of the stars around it. Months later, they took another picture of the same stars, but with the Sun no longer nearby. Comparing the two images showed the deflection of light by the Sun's gravity. Einstein even wrote to his mother when he found out: "Good news today . . . the British expeditions have actually proved the light deflection near the Sun."[56] Because of this bending, it takes the light longer to get to us than if it were traveling in a straight line. A pulsar's companion bends radiation on its way to us—and the additional time it takes for the radiation from a pulsar to get to a telescope is called the Shapiro delay. Using this measurement together with Kepler's laws, it's possible to deduce the orbit of both objects and from that, estimate the masses needed to create those orbits. Timing has been done for years with various telescopes, including Lovell in the United Kingdom, Parkes in Australia, FAST in China, the Green Bank Telescope in the United States, and Arecibo in Puerto Rico.

5

Journey to the Center of a Neutron Star

IT ALL STARTED WITH A GLITCH.

The Vela Pulsar was having a fit. It was early March 1969, and Dick Manchester had just spent a whole day at the Parkes Telescope in Australia. He was observing the pulsar that had been detected only a few months earlier by Alan Vaughan and Michael Large with the Molonglo Telescope. Their find had established the first direct link between neutron stars and supernova remnants. Manchester had been hunting for pulsars for a year now—fresh out of Australia's University of Newcastle and armed with a PhD, he had started his job as a research scientist at Parkes a mere twelve days before the paper about the very first pulsar, PSR B1919+21, or "LGM-1," appeared in *Nature* on February 24, 1968.

Armed with Bell's coordinates, Manchester's own observation of LGM-1 on March 8, 1968, kickstarted his pulsar career—and the recording of its burst of pulses even ended up on the first Australian fifty-dollar note. "I never saw a pulsar that strong again," remembers Manchester, as I meet him in February 2019 in his office at CSIRO in the suburbs of Sydney. His office is small and tidy; rows and rows of physics and astronomy books extend from floor to ceiling. He takes one, titled simply *Pulsars,* which he wrote with Nobel laureate Joe Taylor, and shows me the dedication: "To Jocelyn Bell, without whose perceptiveness and persistence we might not yet have had the pleasure of studying pulsars." He feels very strongly about Bell not having been properly recognized for the discovery, in particular that she did not share the Nobel Prize, and he's not afraid to voice his frustration. "At least in public, Jocelyn is quite philosophical about it," he says. "But the first thing to say is that she discovered pulsars, there's no question about that. Tony Hewish built the array for a different purpose, and she discovered pulsars by accident. But there's no question that it was her insight that led to the discovery."[1]

Not long after Bell's discovery, Manchester entered the pulsar history books himself. On March 2, 1969, astronomer Venkatraman Radhakrishnan had asked him for help setting up Parkes to observe the polarization properties of the emitted radio waves, meaning when the wave is preferentially aligned in one plane. It's probably easiest to understand polarization by thinking about light: electromagnetic waves in unpolarized light, such as those coming from the sun, a lamp, or a campfire, vibrate in more than one plane. Polarized light waves, however, vibrate in a single plane. (Polarized sunglasses use this principle. They are covered in a special laminate pattern that allows through only the vertically polarized light waves and filters out the horizontal vibrations, eliminating glare and making it easier to see the details of any object one is looking at.) Just like light, radio waves are also part of the electromagnetic spectrum; in fact, they work in exactly the same way.[2] Radhakrishnan's observations of the polarization of the radio waves coming from pulsar observations later supported a pulsar model, which describes the emission beam as being directed outward from a magnetic pole.

Manchester was eager to help. The Vela Pulsar was strong, and earlier observations had hinted that its signals were polarized—so the two scientists pointed Parkes at Vela. Suddenly, they noticed a problem: the pulses seemed to be arriving more frequently than expected. In other words, Vela's rotation period of just over eleven times per second, measured in previous observations with great precision by the Parkes team, seemed to be 196 nanoseconds shorter than it was supposed to be. Either the pulsar was now rotating faster, or the telescope was playing a prank on them.

Late in the evening, exhausted, Radhakrishnan retired to bed, leaving his young colleague to figure out what was wrong with Parkes. Manchester spent the entire night checking and rechecking the equipment; he even observed that other pulsars were unchanged. The technology was probably working fine after all. At dawn, he left a note for Radhakrishnan to say that something was up with Vela, not Parkes, and fell asleep himself.

It was indeed Vela that was to blame. By chance, Manchester and Radhakrishnan had spotted for the first time a pulsar "glitch"—in this case, a step change in the period, which is when a pulsar's period suddenly changes by a minuscule amount of two parts in a million. But considering that even back then astronomers could already measure pulsar periods very accurately, that was a lot. And it dropped off very fast, probably in less than a second.

As it happened, a team at the Jet Propulsion Laboratory (JPL) in California, part of NASA, had observed the glitch, too. A few weeks later, two back-to-back papers were published by *Nature*, one by Radhakrishnan and Manchester, and the other by the JPL group. "*Nature* put our article before JPL's—there was a bit of a struggle over who saw the glitch first," says Manchester.[3] It was tricky to determine when exactly the glitch happened, but the JPL observations identified the window as between February 24 and March 3, 1969.

So what caused the glitch? Both teams suggested that the pulsar's moment of inertia had suddenly decreased. This measure indicates how "spread out" the mass of a body is—it is an average of the mass times the radius squared. With Vela, apparently the radius of the pulsar had changed—but how could that be?[4] The glitch was a mystery, and the person who ended up shedding some light on it didn't even know about

Bell's pulsar discovery. In fact, he missed it completely. It was Gordon Baym from the University of Illinois, who in 1968 had been in Japan, working as a visiting professor at the University of Tokyo. To pass the time on his daily subway commute to the office and back to his apartment on the other side of the city, he had been reading a book by Soviet astrophysicist Iosif Shklovsky, *Intelligent Life in the Universe*. Baym was a condensed matter physicist, but Shklovsky's work sparked his interest in astrophysics. When months later he got back to Illinois and found out about pulsars, Baym—together with his colleagues David Pines, Chris Pethick, and Geoff Ravenhall—eagerly started studying neutron stars.

They were theorists, so wanted to get right into the guts of these new objects. They wanted to find out what they might be made of, not least because they all had been studying the behavior of particles and phenomena like superfluidity and dense matter—and these newly discovered neutron stars were as dense as anything could get. Soon Baym would come across Radhakrishnan and Manchester's paper describing the Vela glitch.[5]

BEFORE JOCELYN BELL'S BOMBSHELL DISCOVERY, just a few people had been studying neutron stars. You may remember that in 1939, Oppenheimer and Volkoff calculated the upper mass limit of neutron stars as being 0.7 solar masses, a result that decades later was observationally shown to be way too small. A paper in 1959 by theoretical physicist Alastair Cameron pushed the mass value up to two solar masses, but that was pretty much it—at the time, not many other researchers were interested in objects that would probably never be detected.[6]

Soviet physicist Arkady Migdal was different. He specialized in studying dense matter and especially the nuclei of atoms and was the first to suggest that the atomic nucleus was a tiny analogue for a neutron star. Both are incredibly dense, with matter squished into very little space—although a neutron star is thought to be more than twice as dense as a nucleus. Further, while in an atom all mass is concentrated in the middle, with electrons residing in a cloud around it, in a neutron star it is assumed that the atoms have collapsed, which Migdal back in 1959 suggested would result in a weird state known as a superfluid.

Superfluidity is perhaps the most astonishing everyday manifestation of the realm of the very, very small—quantum mechanics. Typically, in the absence of a current, any fluid will inevitably slow down and stop due to friction—when you pour water on the kitchen counter, it won't move for more than a few seconds. But a superfluid flows forever. How does this happen? Normally, protons and neutrons—also known as nucleons—are very individualistic and tend to avoid each other. At sufficiently low temperatures, however, they form pairs. They start behaving coherently, marching in unison like soldiers; the atoms are now in the same quantum state. This happens only with certain atoms and only when they are chilled to within a whisker of absolute zero. This collective quantum behavior makes a superfluid flow without friction and even climb up walls.[7]

All this was mere theory until 1937. That year, Soviet physicist Pyotr Kapitsa (who later persuaded Stalin to spare Lev Landau) was tinkering with helium, cooling it down to see what would happen. Several years earlier, Kapitsa had been working in Cambridge with Ernest Rutherford at the Cavendish lab, where he had moved after losing his wife and two children to a flu epidemic in Russia. When he traveled to Russia in the summer of 1934, to briefly visit his mother and take part in a symposium, he was—for unexplained reasons—not allowed to return to England. Rutherford tried to help by sending Kapitsa cryogenics equipment from Cambridge, which allowed him to set up a new lab at the Institute of Physical Problems in Moscow. One day in 1937, Kapitsa was measuring the flow of ultra-cold helium into a bath, through a tiny gap of 0.5 microns (an average human hair is 75 microns across) between two glass disks. He found that below 2.17K (−270.98C or just 2.7 degrees above absolute zero)—a temperature dubbed the Lambda point—the liquid flowed almost without friction. Published in *Nature* on January 8, 1938, the work received worldwide attention. "The helium below the Lambda point enters a special state that might be called

Gordon Baym
(Illinois Physics)

a 'superfluid,'" wrote Kapitsa in the paper, coining the term.[8] Landau, who was released from prison in 1939, used Kapitsa's research to develop the mathematical theory explaining superfluidity.

Baym and his colleagues all knew Kapitsa's and Landau's work. They also knew Migdal's theory that the enormous density of a neutron star could turn its core into a superfluid. And they had just learned of the detection of the first four neutron stars, making these objects suddenly very, very real.[9]

As Baym, Pethick, Pines, Ravenhall, and another colleague, Mal Ruderman, learned more and more about pulsars and then read about Manchester's discovery of Vela's glitch, they had a eureka moment. The glitch, they reckoned, was the first evidence that Migdal's predictions had been correct. The cores of these newly discovered, rapidly spinning neutron stars were probably holding a superfluid underneath a solid crust, and a phenomenon called quantum vortices—tiny eddies—might be triggering the glitch. The existence of quantum vortices had been predicted by phys-

Virgo interferometer on the outskirts of Pisa, Italy. This detector, along with the twin Laser Interferometer Gravitational-Wave Observatories (LIGO) in the United States, enabled scientists to determine the exact location of the neutron star merger on August 17, 2017.

(Virgo Collaboration / CCO 1.0 / Science Photo Library)

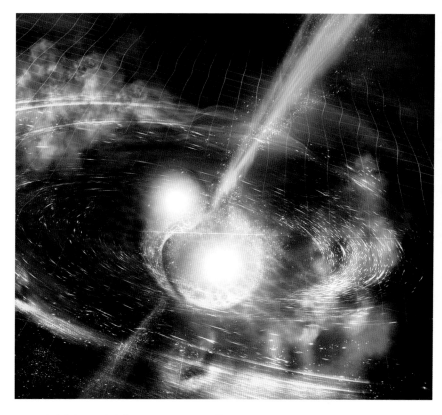

Artist's impression of a neutron star merger. The two neutron stars are tiny—some 20 km (12.4 miles) or so in diameter—but their almighty collision sent ripples across the universe, disturbing space and time. (NSF/LIGO/Sonoma State University/A. Simonnet)

The Parkes Telescope in Australia is one of the veteran radio dishes, completed in 1961. It's famous not only for its tremendous work in pulsar observations, but also for receiving live TV images of the Apollo 11 Moon landing in 1969. (CSIRO)

The author inside the Parkes dish. After I stepped onto the magnificent antenna with operations scientist John Sarkissian, an engineer in the control tower moved the dish with us inside—an absolutely surreal experience. (© Katia Moskvitch)

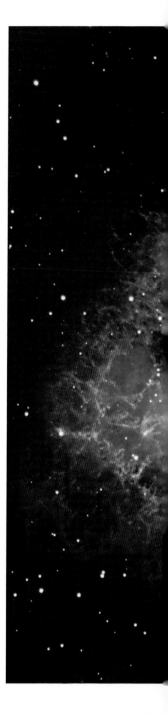

The Crab Nebula is a supernova remnant in the constellation of Taurus. It became famous on November 10, 1968, when astronomer Richard Lovelace and his colleagues discovered a young pulsar lurking inside it, evidence that neutron stars are the collapsed cores of massive stars. (Courtesy of J. Hester and A. Loll / NASA, ESA)

ALMA, the Atacama Large Millimeter/submillimeter Array, is currently the largest radio telescope in the world. Composed of sixty-six antennas, it is situated in the magnificent Atacama desert, 5,000 m (16,000 ft) above sea level. (Courtesy of Clem & Adri Bacri-Normier/ESO)

LOFAR, or the Low-Frequency Array, is a large radio telescope comprising a main base in the Netherlands and several more bases in a number of European countries.

(Netherlands Institute for Radio Astronomy [ASTRON])

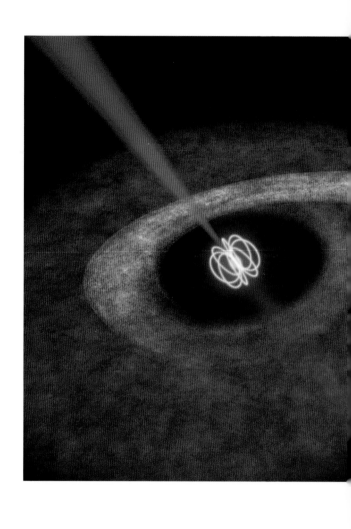

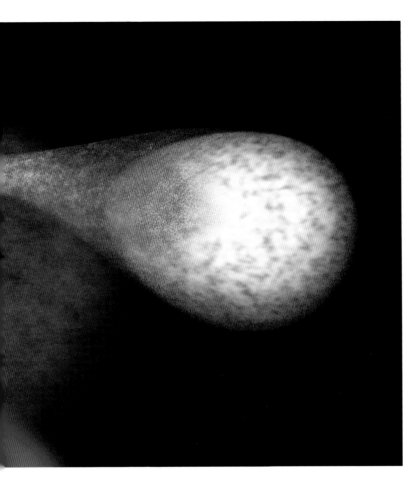

Millisecond pulsars start their lives as regular rotation-powered radio pulsars, but in a binary system with another star, typically a white dwarf. Over time, the pulsar siphons matter from its partner, cannibalizing the other star and speeding up its own rotation to millisecond speeds. (ESA)

The MeerKAT telescope in the Karoo region of Northern Cape Province, some ten hours by car from Cape Town, South Africa, is in a radio quiet zone. It sports sixty-four antennas and is one of the precursors to a much bigger effort, the Square Kilometer Array, which will have many more antennas in South Africa and Australia. (South African Radio Astronomy Observatory [SARAO])

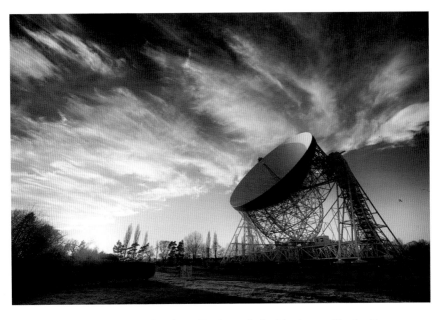

The Lovell Telescope in Jodrell Bank near Manchester, England, has been probing the skies since 1957. It has completed a number of important pulsar surveys, discovering many new pulsars in the process. (Courtesy of Anthony Holloway / University of Manchester)

This image, snapped with MeerKAT telescope, is the best view yet of the center of the Milky Way. While interstellar gas and dust limit optical telescopes, radio waves glide straight through. Sgr A*, just to the right of the image center, is home to our galaxy's central, supermassive black hole. (South African Radio Astronomy Observatory [SARAO])

With a dish 100 m (328 ft) in diameter, the Green Bank Telescope in West Virginia is the largest fully steerable radio telescope in the world. It is this telescope that may one day help us detect millisecond pulsars in our Galactic Center, if they are indeed there.

(Green Bank Observatory)

The Arecibo Observatory in Puerto Rico, built in the early 1960s, is responsible for discovering many pulsars and fast radio bursts. It has also sent messages to space in search of aliens and was the shooting location for two Hollywood blockbusters, *Contact* and *GoldenEye*.
(© Katia Moskvitch)

CHIME (the Canadian Hydrogen Intensity Mapping Experiment) is a radio interferometer in the picturesque wine region of British Columbia, Canada. With four antennas, each sporting huge cylindrical parabolic reflectors, it can capture fast radio bursts by the dozen.

(Courtesy of Andre Renard, Dunlap Institute, CHIME Collaboration)

The South Pole Telescope in Antarctica is part of the Event Horizon Telescope—a planet-sized global network of radio dishes that has given us our first-ever image of the supermassive black hole at the center of Messier 87, one of our neighboring galaxies.

(Courtesy of Jason Gallicchio)

icist Lars Onsager in 1947 when he studied superfluid helium, and further developed by Nobel laureate Phil Anderson and theoretical physicist Richard Packard. "We fleshed out the idea of vortices further," says Baym.[10]

THE IDEA TURNED OUT to be the first observation-based insight into the strange physics of neutron stars. In the five decades since, scientists have still not been able to pin down the inner workings of these ultradense objects. We can't hitch a ride to one and drill into it. So what follows is not science fiction, but it's not quite science fact either. Our problem is that the physics of how matter behaves inside neutron stars is so extreme that our models struggle to truly explain their behavior. But what we know so far is already pretty mind-boggling.

When a proto-neutron star is just born out of a supernova, it's incredibly hot, with internal temperatures soaring to a thousand billion degrees Celsius. A mere minute later, the protons in its interior start morphing into neutrons, producing a huge number of neutrinos. As the neutrinos escape, they carry away energy, causing the star's innards to cool rapidly to about a billion degrees. Its outer layer is already much colder, around half a million degrees Celsius, at which point a solid crust starts to form. Over the next few decades, the neutron star keeps cooling down, its interior reaching a few hundred million degrees, and it continues to slowly lose heat for another hundred thousand years—with the heat inside slowly diffusing to the surface and dissipating in the form of radiation.

One can imagine the neutron star as something like an egg, albeit spherical, with a shell, an egg white, and a yolk. The rigid, crystalline crust is probably about two-thirds of a mile or one kilometer thick and probably made of iron nuclei, the same material that accumulated in the core of the parent star before it went supernova.

Why an iron crust? Because iron is the endpoint of thermonuclear burning: in ordinary stars, over thousands of years hydrogen burns and turns into helium, helium into carbon, and so on; eventually silicon forms, and the ultimate ash of the process is iron. It is not possible to take any more energy out of the iron core of a parent star, hence the assumption that the first outer layer of a neutron star is made of iron. Above this crust there is a thin layer—thought to be between a few inches or millimeters to roughly a meter—of a gaseous atmosphere, whose motion is controlled by the star's magnetic field. The magnetosphere starts just above the atmosphere, and it's these magnetic fields that in rotation-powered pulsars propel jets of particles and hence powerful beams of radiation into space.

A neutron star's crust is an extremely complex structure. As we move toward the star's core, the rapidly increasing density transforms the crust's physical properties. In the outer crust—made from iron crystals—electrons do what we expect them to do: they zoom around the nucleus of each iron atom. As the density increases, however, the electrons gain enough energy to combine with protons. When a negatively charged electron combines with a positively charged proton, the proton morphs into a neutron, releasing a neutrino—and the more electrons get into the nuclei, the richer in neutrons these nuclei become. This process continues as we go deeper, up to the point where there are so many neutrons in the nuclei that neutrons begin to leak out of the nuclei; this marks the transition between the outer and the inner crust, which is called a "neutron drip point," where free neutrons form pairs that make up the neutron superfluid with zero viscosity. This leakage happens from a depth beyond three hundred meters at a density of around 4×10^{11} gm per cm^3: a density that is still less than what researchers have observed on Earth in heavy atomic nuclei. Because of these lab experiments on Earth, the assumptions just noted are based on sound, well-understood nuclear physics. (Though there is no neutron leakage in heavy atomic nuclei on Earth—the nuclei are not sufficiently rich in neutrons for that to happen.)[11]

At the innermost part of the crust, right before the outer core, the density is about a third of the density at the center of an atomic nucleus. The nuclei are now squeezed so close together that, researchers think, they can't stay intact anymore and start changing shape. If before the nuclei were spread out and round like meatballs, now they are all deformed within a sea of dripped neutrons. That's a phase that researchers call "nuclear pasta," with shapes such as tubes, bubbles, and sheets—and hence their names: spaghetti, gnocchi, and lasagna. It's mainly theory, of course, since we can't take samples, but there is very good experimental evidence that nuclei undergo fission—which is the same physical mechanism that leads to the pasta phase.[12]

In the outer core, which stretches to a depth of about nine kilometers (six miles), the density is so high that isolated nuclei can no longer exist. All matter is dissolved into a nuclear goo, a soup of neutrons, protons, electrons, and possibly muons (the heavy cousins of electrons), and with neutrons in a superfluid state similar to ultra-cool superfluids on Earth. Here, though, the temperature is probably millions of degrees—but because the density is so high, it's possible to reach a state of superfluidity regardless.[13]

The insights of Gordon Baym and his colleagues about the Vela glitch have given physicists a hint as to what's happening in the outer core. From their experiments with superfluid helium, researchers now believe that their ideas on quantum vortices were correct and that superfluids don't flow like any other fluid, but instead form those tiny eddies that enable the superfluid's rotation. It is thought that as a neutron star spins down over time—rotating slower and slower—the rotational speed of the solid and liquid components of the star can change, with the crust slowing down faster while the vortices are still stuck in their own local rotation. This would make the superfluid spin slightly more rapidly than the crust and create a lag between the two components. When the lag gets too large, the vortices "jump" to rearrange their rotation—which would make the crust suddenly rotate faster and force the whole star to temporarily increase its rotational speed as well.

That's the idea, anyway—and probably what caused the "glitch" first observed by Dick Manchester back in 1969. After a while, maybe weeks or even months, the superfluid system relaxes and the neutron star returns to the regular speed that we could observe before the glitch. Only about

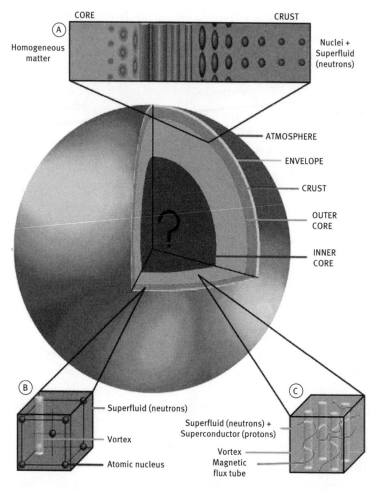

Inside a neutron star, there is a core so dense that it's very difficult to understand what happens there. (Courtesy of Dany P. Page)

5 percent of pulsars have been observed to glitch. Vela is especially curious, because it "glitches" about once every three years. When I visited Parkes in February 2019, John Sarkissian proudly told me that he had observed yet another Vela glitch just days earlier, on February 2. This one and another on December 12, 2016, were particularly interesting, because astronomers found in the data that the pulsar had suddenly slowed down immediately before the glitch—something that had not been observed previously.[14]

As we go deeper, toward the center of the neutron star, things get even weirder and much less certain. Researchers have absolutely no clue as to what's happening inside a neutron star's inner core and the kind of ultra-dense matter we would find there. If we were to understand it one day, the information would give us deep insights into the forces through which particles at such densities can interact. It would also help us to calculate the ultimate mass of a neutron star before gravity overcomes the inner pressure and it collapses into a black hole. At long last, we might understand the final moments of merging neutron stars. But how can we get there?

THE DISCOVERY OF PULSARS has brought together two fields of physics that had been developing in parallel: nuclear physics and astrophysics (using radio, optical, and the very first x-ray observations, which quickly became a key tool for detecting thermal radiation from pulsars' surfaces). The same year the first pulsar was discovered, in 1967, nuclear physicists at the Stanford Linear Accelerator Center started work that a few years later would culminate in their very own breakthrough with the experimental discovery of quarks. Quarks are the fundamental building blocks of matter. Typically, three of them are held together by gluons—glue-like carriers of the strong nuclear force. In such triplets, quarks form regular matter, or baryons—like protons and neutrons that are found in atoms. That's how we encounter quarks today, because they cannot exist freely on their own; instead they have to be confined in baryons.

Confirming their existence motivated physicists to unleash their creativity—and start coming up with various models of the ultra-dense matter in the inner core of neutron stars.[15] Some researchers think that the inner core is composed of mostly neutrons, intact even at such high densities—but this model is "the most boring, vanilla one," says Slavko Bogdanov, an astrophysicist at Columbia University. "There's nothing exotic happening, it's all the same stuff—just neutrons and electrons and protons crushed together in a tiny space."[16]

Other models assume that neutrons do not stay intact but rather break down into their quark components, which would make the core a jelly of free quarks—a quark core. Alternatively, once the quarks are free of the neutron's bonds, they might regroup in other, more exotic configurations—forming particles such as hyperons, where one of the three-quark particles of a neutron becomes a so-called strange quark (normal protons and neutrons have the much more conventional up and down quarks). Another suggestion is that the pressure produces kaons (particles made of two quarks, one of them "strange"), or maybe something completely different. Existing theories on the behavior of quarks and nuclei—such as quantum chromodynamics, or QCD—are helpful, but studying QCD in cold-ish and ultra-dense environments is so complicated that we don't yet have the technology to crunch the results.

To solve the conundrum, researchers resort to what's known as the equation of state—a relationship that describes how the energy density relates to the pressure of the matter inside the inner core—which determines the relationship between the mass and the radius of a neutron star. There are many such relationships, plenty of models of what could be in the inner core, depending on the mass and radius.

Different types of matter respond to gravity's squeeze differently. Imagine that the inner core is a ball—either a dense, stiff baseball that is hard to squeeze or a soft, squishy beach ball. The two have very different behaviors because of the different materials they are made of. For two neutron stars of the same mass, the larger one with a bigger radius would have a stiffer core—because the star is larger, its gravity squeeze will be greater, so the core has to be able to withstand more pressure or it would collapse into a black hole. A baseball-like core would do the trick. But if the core is soft and gets squished easily by the star's gravity, then the star of the same mass has to be smaller in size to counterbalance the squeeze.

Some researchers believe that neutron stars with a stiff core (corresponding to a stiff equation of state) would likely contain intact neutrons, which can be packed only so tightly. Smaller stars with a squishy core (or soft equation of state) would have free quarks in various configurations—not least because processes creating hyperons and kaons out of free quarks would also decrease the pressure. But this is very much subject to debate.

To find out what kind of matter is inside a core, scientists need to calculate how massive the neutron star can be at a given radius. The first step is to observationally measure the radius and the mass of a neutron star, then work backward to get the equation of state—in the process eliminating many incorrect models about the matter inside the core.

The strongest limits on models, or constraints, come from high mass measurements—the higher, the better. Every equation of state has a maximum mass allowed by theory and has to agree with the observations. The heaviest observed neutron star at the moment is PSR J0740+6620, weighing 2.14 solar masses. It was spotted in a Green Bank Telescope survey in 2012. The record-holder before that, PSR J1614−2230, is 1.97 solar masses. Another pulsar of two solar masses was found in 2013. These discoveries killed off theories assuming soft equations of state for particularly squishy neutron stars, because they had predicted that stars of about two solar masses would collapse into a black hole. Among the victims of this constraint were some models that predicted inner cores made from kaons, while others had assumed hyperons inside; scientists spoke of the "hyperon puzzle," wondering whether they had to bid adieu forever to hyperons in the inner core. More recent models, some of which are based on phase transitions of one type of matter into another, still allow for some hyperons to exist.[17]

When LIGO and Virgo detected the neutron star merger, researchers calculated that the upper mass limit of the combined neutron star before it had to collapse into a black hole (which it most likely did) was 2.17 solar masses. The current whopper, PSR J0740+6620, is already very close to that. One issue is that the smallest observed mass of a stellar black hole is larger—some five times the mass of the Sun. That's why some researchers assume that a neutron star that goes above its mass limit might first collapse and become a hypothetical intermediate-mass quark star—but as "hypothetical" in the name suggests, we haven't found any of those yet.[18]

For scientists, all this is like trying to put together a rather complicated jigsaw puzzle; right now, none of their proposed models tell us exactly what might be inside a neutron star's heart. They keep collecting data using telescopes, particle colliders, gravitational wave detectors, and even a special instrument installed on the International Space Station to get new and better measurements of masses and radii. With each set of data, they collect another piece for their puzzle, filling in more of the gaps in our knowledge.

Analyzing Pulsars' "Quark Soup" Core

Sometimes, scientific discovery is rooted in serendipity; like when your surname starts with the same letter as that of a winner of the Nobel Prize for Physics, which means that you share the same university mailbox. Unthinkable in today's world of email, but back then it was standard practice in places like the Niels Bohr Institute at the University of Copenhagen to share big mailboxes, sorted from A to Z.

It was 1970 and Gordon Baym was working at Copenhagen University. Just a year earlier he and his colleagues had written their first astrophysics papers, arguing that the glitch of the Vela Pulsar was observational evidence for a superfluid inside neutron stars. Now he was rooting through the "B" mailbox, checking whether he had received any mail, when he came across a postcard addressed to Hans Bethe. Just three years earlier, in 1967, Bethe had been awarded the Nobel Prize for discovering how stars produce energy. "I turned the postcard over and read it," says Baym.

It was a message from the journal *Astronomy and Astrophysics*, thanking Bethe for a paper he had just submitted on neutron stars. Reading the card taught Baym two things: "One, that Bethe was working on neutron stars. And two, that he was coming to Copenhagen."[19] Excited to meet a Nobel laureate soon, he and his friend and colleague Chris Pethick managed to get hold of a copy of Bethe's paper, hoping to greet him upon arrival with a good understanding of his work. As they worked their way through the paper, spotting mistakes and improving the model, they reached the point where Bethe's theory fell apart completely.

When the Nobel laureate arrived, the pair approached him to break the news that his paper didn't seem to stand up. "So he says, in his wonderful German accent, 'Ve must solve ze problem,'" recalls Baym. "And so there we were, sitting at the foot of the great master of nuclear physics, learning how to do nuclear physics. And that really was a career changing experience."[20] From that moment on, neutron stars and nuclear physics were inseparable for Baym and he decided to understand what's really going on in these tiny, dense, rapidly rotating objects deep in space. He zoomed in on the inner core.

Because many equations of state assume free quarks in different configurations, Baym wondered if it was possible to break nuclei apart to liberate quarks—and then study the resulting "soup," which might be as close to the inner core matter as we can get. Similar conditions existed in the Universe just a few millionths of a second after the Big Bang, when protons and neutrons hadn't formed yet; this baby Universe consisted of a quark-gluon plasma, a weird particle soup.

Baym and other theorists—including James Bjorken and Larry McLerran—had an idea how to set quarks free. The best way, they reasoned, would be to smash heavy ions such as lead or gold nuclei together head-on, at the high speeds achieved in particle colliders. Very sensitive detectors would then register precisely when quarks are created to form a plasma. Even if the quarks would be freed just for a fraction of a second before coalescing into nucleons again, we could gain insights into the possible matter inside a neutron star's heart.

For that, Baym and his colleagues needed a collider. Fortuitously, a failed and abandoned experiment came to their rescue.

It was 1982, and Baym had just been appointed to the Nuclear Science Advisory Committee, a board that reports to the US Department of Energy and the US National Science Foundation. The committee was debating which major facility for nuclear physics to build next, with Baym running a subcommittee on the future of dense matter research. Then, in July 1983, a "miracle happened," at least from Baym's point of view. After more than a decade of construction work, scientists at the Brookhaven Lab in New York realized that the magnets for their planned high-energy proton collider ISABELLE (the name is derived from Intersecting Storage Accelerator plus "belle" for "beauty," as well as, Baym tells me, for a sailboat owned by John Blewett, an accelerator physicist at Brookhaven) simply weren't working.[21]

Rather than seeing ISABELLE through to completion, the scientist wanted to press for a much more capable machine, the Superconducting Super Collider (which was itself cancelled in 1993). With a rival project—the Large Hadron Collider (LHC) at CERN—already under way at the Franco-Swiss border, the US researchers canned ISABELLE, after having spent more than $200 million on the project.[22]

For Baym, the situation was a no-brainer: he was due to give a presentation about the future of dense matter research; instead he now talked about his proposal to use the tunnel to build a different kind of collider, using heavy ions. All the civil engineering was already in place, except for the magnets—so together with a handful of colleagues, he suggested that this would be a wonderful opportunity to actually make an accelerator in Brookhaven. Just like that, he suddenly found himself thinking about building an accelerator—and RHIC, the Relativistic Heavy Ion Collider, was born. Baym was ecstatic. "I was going around giving talks saying RHIC would enable us to figure out the matter of the early Universe—quark matter, before stars and planets formed. But I was also saying that RHIC was going to teach us a lot about neutron stars."[23]

Years passed, and both RHIC and the LHC finally got built. It took until 2000 for the LHC to start getting some results, the same year as

RHIC started colliding its first beams. The LHC smashes lead ions and its collisions result in temperatures as high as 5.5 trillion degrees Celsius, or ten thousand times hotter than the core of the Sun. RHIC, by contrast, orchestrates gold-gold nuclei collisions. When Baym found out in 2003 that RHIC had produced a quark-gluon plasma (the LHC would soon catch up), he was ecstatic. "In a sense, we knew that it had to be there all the time," he says.[24]

It's tricky to visualize this totally new matter that scientists managed to produce. One way would be dividing a second into 10^{23} frames; the substance you get for just one frame is quark-gluon plasma—in an amount so tiny it can fit inside a virus. That's the maximum for how long the quarks stay free, 10^{-23} seconds—after which they quickly coalesce back into protons and neutrons, as well as mesons and other particles.

But there is one problem. The core of a neutron star is much, much colder than the temperatures that RHIC and LHC generate, which are instead close to those right after the Big Bang. Baym knew from the get-go that neither RHIC nor the LHC would ever achieve the much colder temperatures inside a neutron star, which are required for a superfluid to exist. "You simply can't see superfluidity at the heavy ion experiments," says Baym, and sighs: "Temperatures—that's a great difficulty."[25] So instead, researchers are trying to extrapolate their results down to colder levels. In some ways, this is like trying to observe the properties of steam and hoping to deduce the properties of ice. Still, these collisions have a lot in common with neutron star material in the immediate aftermath of a supernova, before the newborn neutron star cools down. And they may help researchers understand what happens when two neutron stars collide, since the remnant in the aftermath of the almighty explosion is really, really hot.

Beyond LHC and RHIC, a new accelerator called the Facility for Antiproton and Ion Research (FAIR) is being built in Darmstadt, Germany—and it may well be able to create quark-gluon plasma at temperatures inside the core of a neutron star. Starting in 2024, its compressed baryonic matter (CBM) experiment will collide high-energy nuclei, pressing them together to create extremely dense matter on a tiny scale—a "fireball." The fireball will then explode, giving birth to about a thousand particles that will decay into electrons, positrons, and muons. The CBM project will study muons because they are not affected by the strong force that holds quarks together, and so may offer clues to how nuclear matter behaves at the high densities found in the cores of neutron stars.[26]

At the moment, though, the best we can get from heavy ion collisions are approximations. We can extrapolate the results to lower temperatures, but such extrapolations are not enough to fully constrain the equation of state and eliminate certain models. For that, we need density and pressure—and hence, a neutron star's mass and radius. One of the ways to measure the mass is with pulsar timing. Numerous telescopes around the world time pulsars, but one of the best next-generation timing machines can be found in a remote part of South Africa's northwest. So that's where I travel next.

Timing Pulsars to Get Their Mass

It's April 2019 and I am in MeerKAT country. And I'm not referring to the small carnivores of the mongoose family, although they roam freely; instead, I'm looking at an other-worldly landscape dominated by sixty-four huge radio dishes, each five stories high. As the sun sets here on the desolate semi desert of the Karoo, about ten hours' drive from Cape Town in the South African province of Northern Cape, the sky is a deep purple all the way to the horizon, but constantly rent by the zig-zags of lightning strikes all around me. MeerKAT—which means "More KAT"—is a follow up to KAT 7, the Karoo Array Telescope of seven antennas. But impressive though it is, it's only the precursor of what is set to become the largest radio telescope ever built on Earth, the Square Kilometre Array (SKA).

To get to MeerKAT from Cape Town one can either fly or drive. Once a week, on Wednesdays, workers, scientists, and the occasional journalist mount a very small plane that usually takes them on a rather bumpy ride. "I've seen so many journalists get violently sick on that plane," laughs

Angus Flowers, the MeerKAT outreach guy who meets me in Cape Town. Because I'm not good with small planes, I opt for a much longer ten-hour journey on four wheels. Before we leave, I am asked to have my audio recorder and camera tested—to make sure they won't emit radio waves that would interfere with MeerKAT's sensitive equipment. I'm told to go inside a "Faraday cage"—a pitch-black room with a thirty-centimeter-thick metal door; a huge propeller spins lazily above to stir up the hot air while I stand there in the dark. After some fifteen minutes my electronics get the all-clear; my gear does not interfere with the frequencies observed by MeerKAT. Phew.

We set off at dawn, leaving Cape Town's famous Table Mountain behind. The beauty of the landscape just beyond the city is stunning, with flat mountains framing the endless orangey-brown plains. Their shape reminds me of a drawing in Antoine de Saint-Exupéry's book *Little Prince*, one of my childhood favorites: the picture seems to show a hat, except it really is supposed to represent a python that has swallowed an elephant. We pass telephone poles struggling to support bird nests the height of a person—built by tiny, sparrow-size weaver birds. We pass through an occasional town, and from time to time I see a farm in the middle of nowhere. Some farms are strikingly beautiful, their large houses surrounded by tall green trees—a true oasis in this desolate landscape. There are no mobile phone towers in the Karoo for miles, and Flowers tells me that many farmers have to rely on satellite phones to communicate in case of emergencies.

About nine hours after setting off, when the sun starts coloring the sky orangey-purple, Flowers turns off the main road and onto a gravel farm track. I spot long, rather unnerving shapes in the road, left by poisonous snakes. An occasional horse stares as the car zooms by. Flowers is in a hurry: he wants to get to the MeerKAT base before nightfall. As we get closer, I take a few more pictures, then it's time for radio and electronics silence. We have to switch off our cell phones, to make sure we don't interfere with the array's highly sensitive antennas as they capture radio waves faraway in space, looking for pulsars and active galactic nuclei—the supermassive black holes thought to live at the hearts of most galaxies. While the telescope has been operational only for a few months, it has already snapped an unprecedented picture of our Milky Way's Galactic Center, showing more detail than any other instrument had ever done, with all the interstellar gas, dust, and weird filamentary threads swirling in there like in some cosmic oven. It's tricky to know what's going on there with optical telescopes because the dust blocks visible light, but radio waves glide straight through. The image is four times the angular size of the Moon (meaning how large the Moon appears when looked at from Earth), and it shows Sagittarius A*, our galaxy's central supermassive black hole.

Now, deep in the Karoo, we pass one last farmhouse and cattle gate, and finally reach a checkpoint. A friendly guard signs us in; he knows Angus, and our visit was approved long ago. After a second security post, we at last approach our destination, a small and seemingly unassuming house.

Inside, I meet Andre, an American engineer from Minnesota who's been working on MeerKAT's data transfer for five years. He's cooking soup with kale and chorizo, which after a day of munching supermarket sandwiches smells absolutely amazing. I'm not allowed to use my laptop or phone in the house either, not until Flowers takes me to the Bunker, an underground building a few meters away ("watch out for snakes and scorpions," he warns me as we walk across). I go through a massive metal door about twenty centimeters thick, which slots in seamlessly. The room is packed with computers, all connected to the internet via an ethernet cable. This is the only place at MeerKAT where any of the workers and visitors can get online.

The next morning, I finally get to see the telescopes. I don ankle-high, steel-capped leather boots—a safety feature against roaming scorpions and snakes, and because this is still a construction site. The MeerKAT antennas are beautiful and in the African sun blindingly white; all fully designed and built locally, they were installed over the previous few months. One, however, looks slightly different. "This is the first SKA dish being installed—a historic moment," says Angus as we drive up for a closer look. Its base is standing like a tree trunk without a canopy; the dish itself is lying on the ground next to it, with engineers in orange hard hats scurrying around. Over the next few years, more than 130 of these antennas will be added to MeerKAT, across an area measuring one square kilometer, hence the name SKA; some 130,000 low frequency dishes will also be installed in a remote region of Australia, part of another SKA precursor, the Australian Square Kilometer Array Pathfinder or ASKAP (see more on ASKAP in Chapter 9).

The story of the SKA goes back to the late 1980s. Just like RHIC, the LHC, and other big projects, these massive observatories take eons to plan, develop, and build. Proposals to build a big telescope array had been bubbling here and there for decades. Canadian astronomers proposed a Radio Schmidt telescope using one hundred twelve-meter (39-ft) antennas, providing a wide field of view; the Dutch wanted to build an extragalactic telescope with a huge collecting area; and the Indians argued that their idea to build an array of 160 dishes, each seventy-five meters (246 ft) in diameter, was the best. There isn't a big international organization for radio astronomy, but scientists did understand that to achieve something this huge, they would have to work together. In 1993, members of the International Union of Radio Science (URSI) took the first concrete step—deciding to establish the Large Telescope Working Group. A few more years went by, filled with debates on the name, the logo, and—most importantly—the site of the future telescope. In the end, the scientists agreed on two spots—in Australia and South Africa, because the Southern Hemisphere offers the best view of the Milky Way Galaxy with the least radio interference. Both countries quickly built precursor telescopes: ASKAP in Australia and MeerKAT in the Karoo.

If all goes to plan, by 2028 the two SKA observatories will scan the sky many times faster than any other radio tools and over a range of frequencies that is so wide, it will be fifty times more sensitive than all other radio observatories put together. The promise for scientists is that they will be able to detect more than ever before, and in much higher resolution, radio sources such as pulsars and other bursts in the sky. Other goals of SKA will be to understand dark energy: a mysterious force thought to be responsible for the expansion of the Universe. The array will also be used to study the formation of the very first stars and galaxies, find out more about the magnetic fields that permeate every nook of the cosmos, and even try to spot other civilizations out there.

The SKA will also be an amazing tool for determining the masses of neutron stars through pulsar timing. (For more on this tool, see "Deeper Dive: Pulsar Timing," in Chapter 4.) It was with the Arecibo telescope that astronomers measured the mass of a neutron star for the very first time.[27]

IT ALL STARTED IN 1974, the same year that Anthony Hewish was awarded a Nobel Prize in Physics for the discovery of pulsars. Astronomers Russell Alan Hulse and Joseph Hooton Taylor were in the Puerto Rican rainforest, in a tiny control room next to the 305-meter (thousand-foot) diameter dish in the ground. They registered a pulse, a new neutron star rotating seventeen times per second. After observing the pulsar for a while, they noticed a bizarre yet regular variation in the arrival time of its pulses; at times they arrived slightly sooner than expected, at times slightly later. A bit of math and reasoning led the two astronomers to conclude that there was a companion, orbiting with the pulsar around their common center of mass. It was the first-ever discovery of a pulsar in a binary system, where a neutron star wasn't isolated but had a companion.

Later, they found out that the companion was a neutron star as well—not a white dwarf, which in later years turned out to be the much more frequent scenario. Hulse and Taylor's observations also led to the very first accurate measurement of a neutron star's mass. They found that their binary's neutron stars had around 1.4 times the mass of the Sun—the pulsar at 1.44 solar masses and the companion neutron star at 1.39 solar masses, both eerily close to the Chandrasekhar limit. As mentioned in Chapter 1, the pulsar now bears its discoverers' name—the Hulse-Taylor pulsar (PSR 1913+16).[28]

So far, astronomers have used pulsar timing to pin down the masses of some thirty-five neutron stars in the range of 1.17 to slightly more than two solar masses. In the Karoo, MeerKAT has started timing pulsars with a project called MeerTIME. The array's observations of known pulsars are already blowing other telescopes "out of the water," says astronomer Fernando Camilo of SARAO (we met him in Chapter 4). MeerKAT and Parkes are both in the Southern Hemisphere, and the Karoo array can focus on those pulsars that have been detected at Parkes over the years, but cannot be seen from the Green Bank Telescope, which is in the north. At Parkes, a neutron star's mass is often calculated with large uncertainties, says Camilo. But MeerKAT, he adds, because of its sensitivity, can make these measurements much better and faster. "Imagine you observe twenty well-known binaries with MeerKat. And within a year or less it detects two or three really well-measured high masses. That could have an enormous impact."[29]

But masses alone are not enough. Researchers are also trying to measure the radii of neutron stars very precisely, to constrain—put limits on—their equation of state. Those are trickier to get, though. A team of researchers led by Michael Kramer, a director of the Max Planck Institute for Radio Astronomy in Germany, has been for years trying to use the data from the only known binary with two pulsars, the Double Pulsar, to constrain the moment of inertia of the neutron star—a combination of mass and radius. Since the masses are already known, it's possible to get some limits on the radius.[30]

But the best size estimates are based on the observations of the brightness of the overall x-ray emission from the surface of pulsars, combined with their distance to Earth. This is not an easy calculation to make, but it has given astronomers radii that agree with the best neutron star theories. Researchers estimate that neutron stars have a radius between 9.9 and 11.2 km (about six to seven miles), although a few also push close to the upper limit with a radius of about 14 km (8.7 miles).

Another way to get the radius is to observe really fast spinning pulsars—millisecond pulsars—and assume that the largest mass that's been measured sets an upper limit on the radius. So far, such a radius upper-limit measurement has only been made once—with the current record-holder in terms of spin.

IT WAS THE FALL OF 2004, and Canadian Jason Hessels, then a graduate student at McGill University, was using the Green Bank Telescope to work on his PhD thesis. The thesis was based in part on Hessels's search for millisecond pulsars in a globular cluster called Terzan 5, some eighteen thousand light-years from Earth in the constellation Sagittarius. He knew that this cluster was home to at least twenty millisecond pulsars—it is one of the most prolific pulsar factories known—and he was keen to find more.

On November 10, 2004, Hessels registered repeating pulses with peaks that were so extremely close together that he realized this neutron star was rotating very, very fast. The problem was that it showed itself only once and disappeared. Days turned into weeks and weeks into months, but it still wasn't there. His colleagues started saying it was just a fluke. But he didn't give up on it, says Hessels, who is now a professor at the University of Amsterdam. He decided that the pulsar had probably simply been eclipsed, hiding behind another star and "behind all this junk in the binary system that has been causing it to become invisible at certain times."[31] Finally, nearly a year later, he detected the pulse again—and it turned out to be the fastest spinning pulsar ever discovered, rotating 716 times every second. It was later called PSR J1748−2446ad. This was significantly faster than the previous record holder, discovered by pulsar astronomer Don Backer in 1982, with a spin of 1.55 milliseconds or nearly 642 rotations every second. Called PSR B1937+21, this was also the first millisecond pulsar to be discovered and was later determined to be around two hundred million years old.

The speed record alone was not the only insight, though. Hessels also managed to determine an upper limit for the radius of this neutron star in a faraway cluster. If the pulsar was any bigger, "we would expect it to be spinning so fast that material would start flinging off its surface," he says. "The pulsar would pull itself apart." He couldn't measure the mass, but assumed the largest mass measured back then, of about two solar masses. Taking the rotation velocity into account, Hessels found that the radius couldn't be larger than 16 km (10 miles). For a more typical mass of 1.4 solar masses, the upper limit of the radius would be 14 km (8.7 miles). Ever since, "it's been kind of a lingering passion of mine to try and find an even faster spinning one," Hessels says—to hopefully also measure the mass and determine the limits of its equation of state.[32]

While pulsar observations have been around for decades, two other very new methods of determining the mass and the radius have just joined it: LIGO and a new instrument that was sent to the ISS in 2017, NICER.

NICER and the Hunt for Hotspots

"Ninjas!" laughed eight-year-old Aeryn when she first saw the picture of a dozen people in blue protective clothing, complete with head masks that leave just a narrow slit for the eyes. Her mother, astrophysicist Anna Watts, says that it's one of her favorite photos, because it shows her colleagues during the final assembly of the best instrument yet designed to tell us what's inside neutron stars. I meet Watts in her office at the University of Amsterdam on my way to LOFAR (the telescope we met in Chapter 4).[33] The Neutron Star Interior Composition Explorer, or NICER, is a 372-kilogram dishwasher-sized box that contains an x-ray telescope. In June 2017, it hitched a ride on a SpaceX Falcon 9 rocket to the International Space Station. Using a robotic arm, astronauts took two days to attach NICER to the ISS.

With NICER, scientists like Watts finally have a tool that should allow them to very accurately measure both the mass and the radius of (nearly) any neutron star that the instrument studies.[34] And on December 12, 2019, NICER released its first results—the best mass and radius measurements of a pulsar.[35]

It took NASA four years to build the instrument. With an array of fifty-six x-ray photon detectors that record the energies and the time of arrival of photons, NICER is designed to study isolated, rotation-powered pulsars. Most of these pulsars emit beams of radio waves generated by particles accelerating in the magnetosphere outside the neutron star. The particles have to return to the surface to keep the star electrically neutral, and this "return current" presumably slams into the surface, heating up the polar caps of the star and triggering the formation of bright areas—hotspots—that radiate x-rays. This is what NICER is after—instead of looking at overall x-ray emission from a neutron star, it documents the exact shape of the pulses from the x-ray glow coming from the hotspots, which depends on the neutron star's mass and radius. That's exactly what Watts had been studying even before NICER was conceived.[36]

In 2015, Watts had visited the lab purely by chance. She was at MIT to give a talk when NICER researchers were testing the detectors. She wanted to have a glimpse of the technology—and being there, had a chance to wander around the lab. The detectors are incredibly delicate, so she had to don the protective gear, a nerve-wracking experience, Watts laughs, "because I'm quite clumsy."[37] A year and a half later, she received an email from NASA, inviting her to join the team. She was excited about joining, but at the time worried whether she could make a useful contribution.

But she did join. Suddenly, her theoretical work on hotspots became much more real. Previous x-ray detectors had already seen these hotspots, but studying them was never the primary focus of any mission.

The idea is to measure how the brightness of the hotspots' x-ray emission changes as a neutron star spins, with the hotspots rotating in and out of view. Because a neutron star is an extremely massive body, general relativity will bend the radiation from the hotspots—to the point that as the star spins, the light from the hotspots on the surface follows the rotation in such highly curved paths that NICER can see the spots even when they are on the back of the star. The instrument tracks how their brightness changes with time, establishing a pattern. It is then possible to accurately predict what an observer will see for a neutron star with a given mass, radius, and spin. "We use this in reverse to infer the mass and radius of the star," says Watts.

Detecting x-ray radiation is tricky, so besides looking for new neutron stars, NICER is also pointing at known radio millisecond pulsars, to try to detect x-ray pulsations from them. If their mass had been previously determined from radio pulsar timing, that's half the job done.

Because at certain times of year the Sun comes near in the sky to these known pulsars, astronomers have to periodically deal with interference from solar photons. "We've actually had to cut some of the data to make sure we only use data taken at certain angles, so that we're not getting this extra contamination from the Sun," says Watts.[38] Even so, the NICER data from observing several nearby millisecond pulsars seem to agree with previous measurements by other telescopes of their overall x-ray emission, which Watts finds encouraging.

A couple of observed pulsars have masses measured from radio timing of more than two solar masses—and for such a large mass, the gravitational light bending is so severe that if the radius is small, the pulses are impossible to detect. So a measurement of the pulsations allows one to place a lower limit on the radius—meaning the minimum the radius can be.[39]

The results released by the NICER team on December 12, 2019, are good, but not yet good enough to determine exactly what the dense matter inside the core might be. The radius is smack in the middle of the band expected from nuclear theory and existing astrophysical constraints, "so the theorists can breathe easy for now," says Watts.[40] But given the uncertainty, and the fact that it's not lying at one of the extremes of the researchers' expectations, scientists can't really make any dramatic statements yet about what it means for the inner core composition.

This may well change for all the next stars the NICER team will be analyzing. While NICER can study only isolated pulsars, future x-ray telescopes such as the enhanced X-ray Timing and Polarimetry (eXTP) mission and the Spectroscopic Time-Resolving Observatory for Broadband Energy X-rays (STROBE-X) will be able to study accreting ones. There, hotspots develop as material from a companion star is channeled onto the magnetic poles of the pulsar, as well as when the neutron star's "oceans" on the surface undergo thermonuclear explosions. While in stand-alone neutron stars the crust is the outermost layer, in accreting ones, syphoned material pools on top of the crust and forms an outer liquid ocean.[41]

Already in planning, these next-generation detectors in space will be ten to twenty times larger than anything that has flown previously and should launch within the next decade. Watts can't wait for this to happen—such a telescope, she says, would allow researchers to pinpoint sharp changes inside the core, like those happening during mysterious phase transitions of quarks (quark matter is thought to change its state just like water changes from liquid to ice to vapor). It's tricky to estimate quark transitions though, because the strong nuclear force that binds quarks together using gluons is still only roughly understood.[42]

LIGO, Virgo, and Their Future Cousins

On the morning of August 17, 2017, when the gravitational ripple of two colliding neutron stars washed over the Earth, Jocelyn Read and Katerina Chatziioannou—both astrophysicists and members of the LIGO collaboration—found themselves in a pickle. Chatziioannou had been woken up by the email alert from LIGO about the collision, which no one had expected to be observed so soon.

Read, a professor at California State University, Fullerton, hadn't even checked her LIGO emails first thing—those emails are sent directly to a special folder she had created because at times the sheer number can seem overwhelming. Both researchers were getting ready to take part in a panel discussion later that day, at an extreme gravity workshop at Montana State University. The subject was whether it was possible to learn anything about a neutron star's interior by observing a merger of two of them.

When she arrived at the workshop, shortly before the panel, Read's LIGO colleagues quickly brought her up to speed. Overcome with surprise and joy, she turned to Chatziioannou. "I had to figure out what I could actually say in the panel discussion without being too obvious," she says—because at that point, the outside world didn't know. The data, however exciting, had to be kept secret and inside the LIGO organization, because it had to be properly analyzed first and papers had to be written.

"It was an unprecedented sort of thing to record, so we were just trying to figure out what to do with it," Read adds.[43]

The panel, in the end, consisted of a lot of "ifs"—but both Read and Chatziioannou knew just how exciting the signal was—a novel way to constrain the equation of state. To do that, all that researchers had at first was the gravitational waves signal, with the hope of charting the frequency evolution of the two stars in their deadly dance. When they are caught in common orbit, they start out orbiting slowly around a common center of mass. As they get closer and closer, over billions of years, they lose energy in the form of gravitational waves; when the two objects fall in together, the orbit speed increases, and eventually the two stars merge. During that time, the frequency changes, and that's what researchers use to extract a lot of information about the neutron star system.[44]

As neutron stars pull closer and closer to each other, they begin to deform—stretching and squeezing the matter on each other and raising tides, just like the Moon's gravity pulls on the oceans on Earth. And as gravitational wave energy leaves the two neutron stars, the tides also take energy away from their orbits, which makes them crash a little bit earlier than they would otherwise.[45]

Despite LIGO's detectors back then not being as good as today, and even though they also captured a lot of noise, the signal was so loud and clear that it was possible to estimate how much the tides—and the dissipated orbital, or gravitational-wave, energy—deforming each star added to the gravitational energy loss. The researchers tried to fit the data into equations of state, to map out how the stars were supporting themselves against the pressure and pull of gravity that would otherwise collapse them.

And even though it's very tricky to predict the final pressures and densities, at the end of the day it's only the pressure-density relationship that determines the size of isolated stars and how the stars interact with tides. The less the deformation, the softer the equation of state that will fit the data, enabling LIGO and Virgo researchers to set an upper limit on the radius from the gravitational wave signal—the radius information is in the small tidal correction to the gravitational-wave energy loss set primarily by the mass of the stars. There are two ways of getting the mass. The most reliable one is to use the so-called chirp mass, or a particular combination of the two masses of the stars, that can be interpreted straight from the early frequency change of the gravitational wave signal. Once additional effects had been disentangled, the masses were found to be 1.46 and 1.27 solar masses, assuming that the neutron stars were spinning with the typical spin of observed pulsars. Read says that it's not a particularly accurate way to determine the component masses, though—the technique is "very promising, but for now the radio pulsar masses are more precise."[46]

Other scientists, among them Samaya Nissanke from the University of Amsterdam, worked on observing the brightness of the kilonova, which can also help with determining the masses, simply by calculating how much mass was ejected during the merger. The observations have put an upper limit to the neutron star maximum mass, of between 2.1 to 2.2 solar masses. But, says Nissanke, the masses obtained this way are "not something I would put any money on" because of the many uncertainties involved—the kilonova, after all, was extremely messy, with lots of material flying around.[47]

Using the gravitational wave signal, Read, Chatziioannou, and others then calculated that the radius of the stars couldn't be more than about 13.5 km (8.4 miles). That agrees with previous x-ray measurements of the radii, and also means that there wasn't much deformation by tides. Such a small radius has helped to rule out the stiffest equations of state that predict larger neutron stars. "That tells us that the pressure is on the low side, so we were able to rule out some of the high pressure models," says Chatziioannou.[48] Had the stars been any larger, the impact of tidal forces on their orbits would have been much stronger, so their internal pressures can't be fluffing them up too much. Yet high pressures are still required to provide a stiff backbone against the crushing gravity of the two solar mass heavyweights, suggesting that there is a strong ramp-up in pressure as densities increase. With more mergers, the next step will be to get much more tidal deformability data. This will allow researchers to probe the radius at different masses and see which equations of state predict the same changes in radius.

Putting the Pieces Together

Once we have precise measurements of the mass and the radius of several neutron stars, the next step is to eliminate the equations of state that don't fit. But even as we pin down the sizes of the stars, that doesn't directly tell us what's causing the pressure: intact neutrons or free quarks, hyperons, kaon condensates, or something even more exotic. In other words, the same measurement of mass and radius can correspond to both a quark interior and a neutron interior; just knowing the pressure doesn't immediately tell us the composition. So the idea is to analyze different scenarios that would produce that type of pressure. That's when puzzle pieces start to fit together.

For instance, by observing more neutron star mergers, it will be possible to look for quark matter changing phases based on varying pressure, hence, how the radius of a star changes with mass. Researchers think that neutron stars in the same mass range will have roughly the same radius. The pressures inside neutron stars are determined by what makes up their matter—which could be plain nucleons, or particles involving strange quarks, or even a soup of quark matter. "It's difficult to determine what particles are generating the radii we measure because of the complexity of their interactions," says Read.[49] But to get radii as small as seen from LIGO data and still allow massive stars, something more "exotic" than the pressure of plain old neutrons may be needed.

And if NICER, for example, detects a lower-mass star with a bigger radius than the current upper limits, and LIGO comes across a higher-mass star with a smaller radius, that would mean that something really interesting might be happening with the particles in the core—telling us a lot about what kind of matter is producing the pressures involved and constraining the models. "If those observations are giving us two different views, two different types of neutron stars, that could enable us to not only precisely map how the equation of state behaves but also what particle or fundamental interaction causes it to behave like that," says Chatziioannou.[50]

We could eliminate additional models by combining our knowledge of phase transitions, cooling rates, how different particles radiate away energy, and how neutron stars spin down over time. Future collider experiments of quark-gluon plasma meanwhile could give us clearer ideas of what's inside these dense bodies.

But will we ever be able to eliminate all but one theory? Chatziioannou is not so sure. "If we know theorists, that's unlikely—there will always be new theoretical models that fit the data, especially if our observations are challenging the existing understanding. You can maybe eliminate extreme phenomena such as really large quark cores, but there will always be possibilities where there's a small amount of quarks. There's no way you can completely eliminate everything."[51]

• 6 •

How Neutron Stars Keep Spoiling Dark Matter Theories

GAS, DUST, MILLIONS UPON MILLIONS OF STARS, a handful of pulsars, and a monstrous black hole a million times more massive than our Sun. That's the Galactic Center, the beating heart of our Milky Way, around which all the stars and planets have been rotating for some 13.5 billion years. It's a very crowded place. So crowded that it's impossible to peer into it with optical telescopes: the wavelengths of visible light simply can't penetrate all the gas and dust. Luckily, long radio and infrared waves glide through, revealing the gas and old red supergiant stars, as well as a few million younger massive stars that will go supernova many millions of years from now.[1]

In June 2018, ahead of its official inauguration on July 13, the MeerKAT radio array in South Africa briefly glanced at the Galactic Center and snapped a picture of it. The result was the most beautiful, closest, and sharpest image of the heart of the Milky Way ever captured. It looks like a bonfire or a raging oven. Flame-like filaments stretch in all directions, in clear lines. Weird round clouds on the left-hand side evoke a puffing chimney.

Two huge radio bubbles extend up and down in an hourglass shape. Hiding bang in the center is the supermassive black hole Sagittarius A*.[2]

But there's more. Astronomers think that the center is hiding another secret—matter completely invisible to any of our instruments. This peculiar substance is thought to amount to more than five times the mass of the mundane, familiar atomic matter that scientists call baryons and that we all know about—the matter that makes up stars, gas, planets, you, me, and Quark, my cat. We don't know what this matter is, but as Caltech's Fritz Zwicky first theorized, we need it to explain the observed motion of stars and galaxies. We now call it simply "dark matter."

About 23 percent of the Universe is made up of dark matter, 72 percent is composed of the mysterious dark energy, and just under 5 percent is normal matter. There's just one problem: it's now been a few times that, when scientists think they may have spotted a clue to the identity of dark matter, they find themselves facing the same old foe that enters the scientific arena and undoes their theories—the neutron star.

Most astronomers believe that they have enough indirect evidence for dark matter. They see its gravitational pull on stars, they see it bend light passing next to it. It all adds up, but only if we assume that Albert Einstein's theory of general relativity correctly defines gravity as a consequence of massive objects warping spacetime.[3] (There are alternative theories of gravity that try to explain our Universe without resorting to dark matter or dark energy—but these theories are fighting their own battles with neutron stars. For more on this, see Chapter 8.)

Still, if general relativity is correct, then we simply can't explain the gravitational movement of the observable normal matter without resorting to dark matter. As with many seemingly far-out theories, it all goes back to the guy who back in 1933 first linked neutron stars and supernovas—Zwicky. At around the same time, he was studying galaxies and galaxy clusters, like our very own cluster, the Local Group that includes the Milky Way, Andromeda, and many smaller, dwarf galaxies. Zwicky knew that galaxies in a cluster rotate around the cluster's center of mass, as well as rotate around and interact with each other.

But he wanted to have a look for himself. He persuaded Caltech to build an eighteen-inch telescope with a wide field of view and took photos of the Coma cluster, some 323 million light-years from Earth. This cluster contains more than a thousand galaxies. Each image that Zwicky took had a large number of galaxies, and he counted the luminous matter (stars and gas that he could actually see through the telescope) in the cluster. He also calculated the velocity of the galaxies—and realized that the cluster was much more massive than one would have expected from just adding up its luminous matter. That was startling. The galaxies seemed to move way too fast to actually stay gravitationally bound inside the cluster. With little luminous matter to provide mass and gravity, the Coma cluster should have simply flown apart. Instead, the observable gravitational forces suggested the presence of four hundred times more matter—which he called *Dunkle Materie,* or, translated from German, dark matter. Not many people paid attention to Zwicky's theory until decades later, but the name survived.

It wasn't until the 1970s that scientists realized Zwicky had been right all along. Two American astronomers, Vera Rubin and Kent Ford, had been calculating the mass of our neighboring Andromeda galaxy by measuring the velocity of its stars. Similar to Zwicky with the Coma cluster, they found an anomaly. Andromeda is a spiral galaxy, like the Milky Way, with a galactic bulge and a disk full of interstellar gas and stars. Before Rubin's work, astronomers had assumed that stars rotate faster closer to their galactic center, while the spiral arms farther out would move slower, given the expected weaker gravity. Rubin, however, realized that this wasn't the case; stars had the same velocity on the outskirts and closer to the heart of Andromeda. This meant that the galaxy had to be filled with something invisible that had enough gravitational effect to hold it all together—some mysterious "dark matter" that was helping normal matter to spin at the same speed across the whole galaxy. Rubin and Ford later performed the same calculations with many other galaxies, and by showing that the flat rotation curves were ubiquitous in galaxies, they made a convincing argument for dark matter.

That was the turning point in the debate over this invisible stuff. Astronomers now believe that dark matter forms big, puffy, spherical clouds throughout the Universe, with one or more galaxies floating inside each one, like futuristic cities in the sky—called dark matter halos. Later, astronomers also saw the evidence for dark matter's existence in the bending of light in space, an effect called gravitational lensing. When we observe a faraway galaxy, often the image we see looks distorted, like an arc, or it may even appear as multiple images when we know there's only one object there. That's because between the object we are observing and the Earth could lie, say, a galaxy cluster, which acts like a giant gravitational lens, bending the light rays around it. But when we add up all the stellar mass in the cluster according to the amount of light we detect, we calculate that the amount of lensing should be negligible—the image of the distant galaxy shouldn't be that distorted at all. When astronomers try to recreate the distortion in computer simulations, they often have to input as much as nine times more matter into the simulated cluster between us and the object; only then does everything work out. This suggests that the cluster contains much more dark matter than normal matter.

Despite spending billions of dollars and decades of work on various detectors, scientists have been unable to find any direct evidence of dark matter, and attempts to reproduce it during high-energy collisions in particle accelerators such as the Large Hadron Collider at CERN have yielded no results so far.

On one of my travels, I end up deep under the Italian Appenine Mountains, about an hour's drive from Rome. Here, a lab is hidden deep below the highest peak, Gran Sasso. In the lab, I'm shown a giant tank about 10m in diameter and 11m tall, reminding me of a large grain silo. This is the most sensitive dark matter detector on Earth, XENON1T. But despite decades of searches, scientists here—or at any other detector—haven't found the signal they've been looking for.

Still, astronomers aren't giving up; based on the observed gravitational effects, their assumption is that the densest concentration of dark matter around a galaxy or a group of galaxies should be in the messy Galactic Center. That's because according to the accepted models of galaxy formation, in the early Universe dark matter halos formed first, when this substance gravitationally collapsed into a web-like structure called the cosmic web, with blobs of dark matter at the nodes. Those blobs are the halos, huge American football–shaped clouds of dark matter—with denser concentrations of it in galaxies' center. So if you want to find dark matter, they think the best place to look for a signal should be the Galactic Center.[4]

DAN HOOPER, an astrophysicist at the University of Chicago, knows that full well. For two decades now, he has been hunting for dark matter—searching high and low across the Milky Way for any signals that might prove its existence. Hooper has been zooming in on the Galactic Center and also examining the Galactic Plane—where most of our disk-shaped galaxy's mass is found. But every time he thought he might be onto even the slightest hint of dark matter, his search has been thwarted—by pulsars, which happen to produce signals similar to the presumed dark matter signals that Hooper has been searching for. "Pulsars are kind of the foil to dark matter," he laughs. "When you observe something exotic that you think might be dark matter, you worry that it's actually pulsars."[5]

The story of the battle between pulsars and dark matter starts in the late 1990s, when a ground-based detector in New Mexico called Milagro observed a diffuse emission of extremely energetic gamma rays from across the Galactic Plane. Gamma radiation is the most energetic form of radiation we know, and Milagro registered its upper limit, of energies around 3.5 tera-electron-volts (TeV), or one trillion electron-volts (eV). The eV describes how much energy an electron, moving at a third of the speed of light, gains after being accelerated by one volt of electricity. For comparison, in old cathode-ray televisions electrons were accelerated by about thirty thousand volts, meaning that when they hit the screen, they carried the energy of thirty thousand eV. So 3.5 TeV would be about a hundred million times more energy than that—bearing in mind that a 3.5 TeV electron travels much faster than those in cathode-ray TVs, zooming at a whisker below the speed of light because it's so much more energetic.

Dan Hooper

(Courtesy of Dan Hooper)

The fact that there was gamma radiation across the Galactic Plane wasn't a surprise. Gamma rays are produced during the radioactive decay of atomic nuclei. On Earth, they are born in nuclear explosions and lightning bolts. In space, they are generated by cosmic rays—usually high-energy protons—when they interact with interstellar gas, or when electrons zip through galactic radiation fields or starlight, and in highly energetic events and objects, such as supernova explosions, regions around black holes, and neutron stars. But when researchers analyzing Milagro's data compared their findings with how much gamma radiation was supposed to emanate from the Galactic Plane based on the amount of cosmic rays observed, they were shocked to find at least ten times more of this radiation than there was supposed to be.

Then, in 2006, the space-based experiment called PAMELA (Payload for Antimatter Matter Exploration and Light-Nuclei Astrophysics), on board a Russian satellite, detected another excess—this time, an odd spike of high-energy positrons, the anti-particles of electrons. Scientists believe that the Big Bang should have produced equal amounts of matter and antimatter. Somehow, though, matter came to dominate, and very little antimatter is found in the Universe today—although it is possible to create it, even on Earth. The galactic magnetic field randomizes the motion of these particles, so it was impossible to pinpoint the source or sources of these positrons. But Hooper knew that electron-positron pairs scatter the ambient starlight up to ultra-high gamma-ray energies at the TeV level. So were these two signals linked in any way? And what was generating them?

When PAMELA's data suggesting a positron excess were made public, Hooper was one of the first scientists to react. His first thought was that the excess might be the sign of what he had been searching for all his life: dark matter. Scientists have come up with a number of theories for what dark matter might be, and one of the most widely supported concepts is that it is made up of WIMPs—weakly interacting massive particles. One type of WIMP is a hypothetical neutralino, and Hooper suggested that these neutralinos might occasionally collide, and as they were annihilated, produce showers of exotic particles that then decay into regular elementary particles—electrons and positrons, leading to the observed positron excess. According to this view, these electrons and positrons, moving at nearly the speed of light, then spiral around the galactic magnetic field, generating what's known as synchrotron radiation (emitting electromagnetic energy); the electrons also scatter low-energy photons of ambient starlight to very high gamma-ray energies via a process called inverse Compton scattering.

Hooper wanted to think it was dark matter, but couldn't be sure, so he also considered an alternative explanation—that the positron excess might be due to pulsars. Their powerful magnetic fields, spinning with the neutron star, generate an electric field that drags electrons from the surface of the pulsar and accelerates them. As these highly energetic electrons travel through the magnetic fields, they emit, just like in the dark matter scenario, high-energy gamma radiation—and some of this radiation will spontaneously change into electron-positron pairs as it escapes the pulsar's magnetism and zooms across space.

Other physicists were as intrigued as Hooper; within a decade of PAMELA's data becoming public, around a thousand papers had been published trying to explain the mystery. The majority of papers favored the pulsar explanation, but the challenge was to actually detect these gamma rays coming from a pulsar, then determine how many electrons and positrons it would take to produce this amount of highly energetic gamma radiation, and whether it might match the observed excesses.

In 2017, Hooper and a handful of colleagues came up with an idea for how to solve the mystery. They turned to yet another detector—Milagro's successor, the High-Altitude Water Cherenkov Observatory (HAWC) near Puebla, Mexico, which was completed in 2015. HAWC had been observing two nearby pulsars, Geminga and the Monogem ring pulsar (some call it simply Monogem), located less than a thousand light-years from Earth. Their relative proximity to us is important, because electrons don't travel very far before they lose much of their energy in the magnetic fields of our galaxy, and because of the scattering of starlight.

HAWC observed that the two pulsars had a broad gamma-ray halo in the highly energetic TeV range. These gamma rays could be generated when high energy electrons and positrons, produced by the pulsars, interact with low-energy photons from nearby stars. The collisions transfer a lot of energy to the photons, like a golf club sending a ball to the other side of the field. The researchers analyzed the HAWC data and calculated the luminosity of its two sources. They compared the brightness of the two pulsars to the brightness of their halos and determined how much of the pulsar energy was being converted to electrons and positrons. It turned out to be roughly 10 percent—and that, says Hooper, was pretty much what was needed to explain the positron excess. It was, as he says, the smoking gun evidence.

Nearly three thousand radio pulsars have been detected in our galaxy, although most of them are either too dim or too distant for HAWC to distinguish as individual sources. But if one assumes that all pulsars are equally efficient at converting their kinetic energy into electrons and positrons, adding up the contribution from all the pulsars in the Milky Way, that would produce a TeV gamma-ray signal with almost exactly the right intensity and spectrum (the range of all types of electromagnetic radiation) as the observed TeV excess. Nearly everyone now agrees that the positron surplus is likely due to pulsars, and the original, competing dark matter theory has been shelved. Pulsars have won this fight.[6]

HOOPER WAS DISAPPOINTED, but not crushed. While studying the weird signals from the Galactic Plane, he was searching for dark matter elsewhere—pinning most of his hopes on the heart of our galaxy, the Galactic Center.

Back in 2003, a young postdoc at Princeton, Doug Finkbeiner, was combing through data from a space-based satellite called Wilkinson Microwave Anisotropy Probe (WMAP). Just like Hooper, he was fascinated by dark matter—and hoped to stumble upon signs of it in the WMAP data. He knew full well that WIMPs, if they exist, should occasionally annihilate and in the process produce a cascade of gamma rays, microwave radiation, and other high-energy particles. He found a weird surplus of microwaves around the Galactic Center. It was clearly different from the primordial cosmic microwave background, because it seemed to originate somewhere within our galaxy, meaning that it was more foreground than background radiation. Finkbeiner's paper on the peculiar microwave haze caught Hooper's attention—and especially the suggestion that WIMPs might be the culprits behind it.

The picture got even muddier in 2008, when NASA's Fermi Gamma-ray Space Telescope returned its first batch of data. Finkbeiner, now a professor, and his two graduate students, Tracy Slatyer and Meng Su, found that the microwave haze was complemented perfectly by a gamma-ray haze around the heart of the Milky Way. But where this haze was coming from was a big mystery. Unlike radio telescopes, which target a specific area of the sky and register a signal from a point source—like a star or a pulsar—Fermi has a much wider field of view. It sees about a fifth of the entire sky at any given time, sweeping through the cosmos to capture the whole sky every three hours. Still, its vision is blurry—it has an angular resolution of only about 0.1 to 1 degree at most energies, so it smears out stars to the size of the Moon or larger. So despite being able to determine photon directions, Fermi couldn't see where the haze, later called the Fermi haze, was originating exactly. All researchers were certain about was that it happened all around the Galactic Center.

Shortly thereafter, in 2009, Slatyer, Su, and Finkbeiner noticed that the haze had edges. This discovery was surprising and unexpected—instead of seeing just a fuzzy blob, suddenly they were looking at two gigantic, mesmerizing bubbles extending in an hourglass shape some fifty thousand light-years from top to bottom, centered on the Galactic Center, and shining most brightly in the gamma-ray spectrum. The Fermi haze became Fermi bubbles, and a stunning image of a huge figure eight in space made headlines around the world. Slatyer, Su, and Finkbeiner bagged the 2014 Bruno Rossi Prize of the American Astronomical Society's High Energy Astrophysics division for the find.[7]

For Hooper, the bubbles were a disappointing but clear sign that neither WMAP's microwave nor Fermi's gamma glow could be a dark matter signal. "Dark matter annihilation shouldn't make these kinds of bubbles," he says; instead it should be smooth and without borders.[8] While it's still unclear what blew up the bubbles, some researchers think they could be due to an ancient outburst of our supermassive black hole, even though Sagittarius A* is totally calm now, unlike the active galactic nuclei (AGNs) that we observe in other galaxies and that emit radiation. An alternative theory suggests that the bubbles were formed when numerous giant stars, born from all the gas around the black hole, exploded nearly simultaneously as supernovas.

But Hooper was tenacious. As soon as he learned the haze couldn't be dark matter, he decided to reexamine the same batch of Fermi data, together with New York University grad student Lisa Goodenough. And in that data they found a hidden gem. They saw that around the Galactic Center there seemed to be much more gamma radiation than theory suggested; it was more than all the known astrophysical sources taken together—the supermassive black hole, cosmic ray protons striking gas, cosmic ray electrons scattering off photons and gas, as well as supernova remnants and the Fermi bubbles—could possibly emit. The gamma radiation was running at a surplus of about 10 percent. On top of that, the spectrum and distribution of the observed emission didn't match what researchers had expected to see from purely astrophysical sources. Hooper was stunned. Had he found evidence for dark matter at last?[9]

The excess radiation seemed to be most prominent at energies between one and three billion electron-volts (gigaelectron-volts, or GeV), which is about a billion times more energetic than visible light. It was brightest at the Galactic Center, and became fainter the farther away it was from the center, in any direction—extending roughly spherically at least five thousand light-years beyond the center. Hooper suggested that the excess was due to annihilated WIMPs—and with Goodenough, he posted a paper on arXiv.org, a popular online platform hosted by Cornell University that makes it easy to share scientific papers before they appear in peer-reviewed journals.[10]

But no one seemed to take notice.

Hooper tried to accelerate matters. At a conference that year he gave a talk about his team's calculations. He recalls how he enthusiastically suggested that there might at last be a sign of dark matter—in the Galactic Center. But the audience was eerily silent. After the talk, one researcher walked up to Hooper and called him an amateur who simply didn't know how to properly interpret Fermi results. Similar negative reactions followed at other conferences, especially from members of the Fermi collaboration, who were the primary users of the telescope's data. Hooper was an outsider, after all, simply dissecting the results that the Fermi team had made public. "They were very dismissive," he sighs.[11]

A few months later, when Fermi delivered another batch of data, Hooper and Goodenough saw that the unexplained bump of radiation was still there—for all to see. "We took a lot of the criticism and tried to do a better version of the analysis, including much more sophisticated treatment of how the instrument works, and more sophisticated treatment of the backgrounds and point sources in that field of the sky. And we just kept finding the same sort of signal; it just wouldn't go away," says Hooper. They published a second paper in 2010. But again, not many people paid attention.[12]

For Hooper, though, this paper was a turning point because he was now certain that the signal was real. He wasn't sure whether its source was dark matter annihilation, but the signal was clearly there.[13] In September 2010, a young grad student from the University of California, Santa Cruz, came to work with Hooper. His name was Tim Linden, and he had heard one of Hooper's early talks. Now an astrophysicist at Ohio State University, Linden recalls the hostility faced by Hooper: "To say that the Fermi collaboration was not taking Dan seriously is an understatement; there were some people that were very dismissive of the analysis for years. And I always thought there was nothing wrong with the analysis." Was it possible that there was some other source that might explain the gamma rays? "Yeah, I think that's always a possibility. It's still a possibility now," says Linden. But still, he was intrigued by Hooper's claims—and convinced that the astrophysicist was on to something.[14]

The two published a paper in October 2011 describing how they had found the same amount of excess, even though they had used a different technique to measure it—they had made maps of the sky in the gamma-ray spectrum. Other teams of scientists finally followed suit and started finding too much gamma radiation as well. Those working on the Fermi collaboration were still hard to convince, though, and many researchers were adamant that the bump was not real. It was time, Hooper thought, to write the definitive paper about the excess—and either make the case or bury it.[15]

So in 2012 Hooper reached out to some of his colleagues working on gamma-ray analysis, including Finkbeiner and Slatyer, who was now a postdoc at the Institute for Advanced Study in Princeton, New Jersey.

Slatyer was still exploring the origin of the bubbles. One summer day that year, at a conference in Aspen, Colorado, she bumped into Hooper and they started talking about what was happening to the bubbles closer to the Galactic Center. Not many people thought by that time that the bubbles could be a sign of dark matter, but claims of similarity between the bubbles' light in microwaves and gamma rays were based on comparing the gamma rays far away from the Galactic Plane to the microwaves closer to the Galactic Center. Since the comparison wasn't apples-to-apples, Hooper suggested to Slatyer that maybe the inner part of the bubbles could be associated with dark matter, after all. "We discussed whether the part of the bubbles closest to the Galactic Center was being 'contaminated' by the Galactic Center excess—which I was arguing could be from dark matter," says Hooper.[16]

Slatyer listened, intrigued. She knew about the gamma-ray excess—and agreed to examine the bubbles, joining Hooper's search by zooming in on the Galactic Center. Over the next couple of years, she and Hooper published two big papers together—the second one jointly with Linden and Finkbeiner, a thirty-page-long work that took half a year to produce.[17] They double- and triple-checked everything, meticulously addressing every potential criticism. "It was kind of a monumental effort, it was full-time for six months," recalls Linden.[18]

No matter which models they ran or how they processed the data, the bump—the excess in gamma radiation—was still there. They even deliberately looked at different parts separately, with each team member tackling their own bit of the problem, to avoid introducing any bias. "I was working on the Galactic Center part of the analysis and was totally blind to what the people doing the inner galaxy were finding. And Tracy and her group were working on the inner galaxy part of the analysis, and they were blind to what I was finding; and then we kind of matched it up," says Linden. They found that even at a fairly large distance from the Galactic Center, several thousand light-years out, the excess was present.[19] "Nobody had previously realized that it wasn't just at the Galactic Center, but extended all the way into the inner galaxy," says Slatyer. That was interesting, because it immediately eliminated concerns that the glow might just be the emission from where the black hole was gorging on surrounding matter. "Those explanations just didn't really work anymore, once you saw that this emission appeared to extend thousands of light-years away from the black hole," she says.[20]

Their paper also showed that the "cloud" of gamma radiation was roughly spherically symmetric around the Galactic Center, and that its energy spectrum looked extremely similar everywhere, even far from the heart of the galaxy. As they began to piece together their findings, they realized that they were probably looking at a large number of sources—either dark matter or something else that was distributed broadly symmetrically around the center of the galaxy.

The paper appeared on the online pre-print server arXiv in February 2014, with the title "The Characterization of the Gamma-ray Signal from the Central Milky Way: A Case for Annihilating Dark Matter"—and immediately generated a ton of press, says Linden. (It was finally published in a journal two years later, in June 2016.[21]) It was the first paper about the mysterious glow that journalists would be covering daily, for weeks. The Fermi collaboration could no longer ignore the unexplained surplus and even NASA put out a press release, on its website's front page, acknowledging that the excess existed. "It only stayed there for about twenty minutes before they changed it to something else, but I managed to send the link to my mom," laughs Linden.[22] The work sparked other papers, with researchers citing it more than 150 times in the following three months. A year later, even the Fermi collaboration published a paper on it. At last, the debate had shifted from whether any excess existed to what could be the radiation's source.

Some researchers sided with Hooper and his dark matter theory. But not all. And that's when pulsars struck again—because many scientists preferred the much more familiar explanation of the glow, and this possibility had been raised by scientists (including by Hooper) as early as 2010.[23]

Stars at the center of the Milky Way tend to be a lot older than those at its outskirts. The oldest stars in our galaxy are about 13.5 billion years old—so for a very long time, massive stars at the center should have been going supernova, giving birth to a lot of neutron stars. Because the density of stars in the Galactic Center is incredibly high, these neutron stars could have been forming in binary systems—meaning that at some point they'd be siphoning matter off their companion and hence rotate extremely fast.[24] They would have turned into old recycled millisecond pulsars—emitting gamma rays and possibly radio waves. Based on the gamma-ray observations of known millisecond pulsars, researchers calculated that such an excess could be produced if there were several thousand pulsars in the Galactic Center. And while they are point sources, if they are clustered together, their spectrum would be very similar to the one that might be generated by the annihilation of WIMPs. But when Linden and Hooper scanned catalogs for known gamma-ray pulsars in that region, they didn't find any. And, Hooper says, how likely is it anyway that thousands of pulsars are hiding at the heart of our galaxy, unbeknown to us? Why are we not seeing them?[25]

Part of the answer to Hooper's question may be related to technology. If you take the brightest millisecond pulsars that we've ever detected and put them at the Galactic Center, it would still be tricky to spot them. "We're right on the edge of being able to detect them," says Slatyer.[26] It's not that astronomers haven't tried to find them—they have, but until MeerKAT came online and snapped its picture of the Galactic Center, one of the most sensitive telescopes searching for millisecond pulsars in the region had been the Green Bank Telescope in West Virginia.

So that's where I decide to travel next—because it is this observatory that might at long last settle the long debate over pulsars versus dark matter in the Galactic Center.[27]

AS I DRIVE ON A FOREST ROAD through Pocahontas County, West Virginia, four hours from Washington, DC, I am experiencing digital detox. For about an hour now, I've been in the National Radio Quiet Zone—a unique region in the United States, spanning about 21,000 square kilometers (13,000 square miles), where cell phones don't chirp and radio transmissions are highly restricted by law.[28] After roughly two hours of driving on snaking roads through the Blue Ridge Mountains, I arrive at the tiny town of Green Bank, population 143. A few cows idly gaze at me from a field next to the road; a guy in a checkered shirt repairing his pickup truck on the curb looks up and nods, smiling. I turn into a small parking lot surrounded by low buildings.

Behind these buildings, reaching an impressive 148 meters (485 ft) high—more than twice the size of the Statue of Liberty—towers the instrument for which phones, WiFi, and even microwaves in the area have been banned: the magnificent Robert C. Byrd Green Bank Telescope, or GBT, a metal structure weighing 7,600 metric (8,378 imperial) tons. It's visible for miles from the surrounding country roads. Locals call it the Great Big Thing. Within a 32 km (20 mile) radius around it, the Green Bank Observatory constantly scouts the area for devices emitting high amounts of electromagnetic radiation—and when the operators find one, they ask the violator to stop using it. They can't legally demand that a citizen throw away a microwave or a router—but they do try to reason with the person to find a solution.

The telescope is painted snow white every year, with more than 5,000 liters (1,320 gallons) of paint needed for three coats. All radio telescopes are typically white to reflect sunlight and heat, minimizing temperature fluctuations of the surface of the antenna and keeping it from warping. GBT started taking data in 2001, scanning the sky for radio waves with frequencies ranging from 100 MHz to 100 GHz. Back then, it belonged to the National Radio Astronomy Observatory (NRAO). In 2012, though, the NRAO decided to stop funding the facility, citing budget cuts. The telescope was scheduled to shut down on October 1, 2016. But instead of letting it die a quiet death, its team decided to turn the observatory into a self-funded facility. Its dish is a hundred meters (328 ft) in diameter, making it the largest steerable antenna in the world, and it searches for pulsars at frequencies below two GHz; apart from China's FAST and Arecibo, it is the most sensitive single dish in the world.

While GBT itself is not even two decades old, it is surrounded by several much older telescopes. When I'm taken on a tour of the grounds, wearing a red hard hat, with my phone in airplane mode and with special permission to take digital photographs (usually only cameras with old-fashioned 35 mm film are allowed), I see GBT from afar—and it dwarfs all other telescopes on the site, old or new. First, I get to see the rusting Howard E. Tatel telescope—the very first radio telescope of the NRAO—which started observations on February 13, 1959. Thirty meters (85 ft) in diameter, it shot to fame in 1960 when the renowned astronomer Frank Drake used it to for the first time for SETI, or Search for Extraterrestrial Intelligence. The two-month long initiative, called Project Ozma (after a character in the book series the *Land of Oz,* by L. Frank Baum), observed two of our nearby stars, Tau Ceti and Epsilon Eridani. It didn't find aliens—and one briefly detected signal later turned out to have come from an airplane flying past. Still, the veteran telescope did make a number of discoveries, such as documenting more precise locations and brightness of a number of radio objects, measuring the surface temperature of the Moon and Venus, and studying the radiation belt of Jupiter. Later, in the mid-1960s, two identical new telescopes at the observatory joined forces with the Tatel to become the Green Bank Interferometer (GBI). Then a few smaller, portable telescopes were added to GBI as well—turning it into an array prototype for the Karl G. Jansky Very Large Array in New Mexico. The GBI was first to confirm that massive bodies in space bend light—testing Einstein's theory of general relativity.[29]

The design of the GBT itself is slightly different from that of Parkes and Lovell. Radio waves from space are reflected to the telescope's extremely sensitive receivers at the top of a side structure, the boom, which is placed to one side so that it doesn't obstruct the radiation arriving at the dish itself. This side boom looks like a muscular arm held up by the mighty instrument as it shows off its strength in an endless moment of triumph. And there's a lot this telescope can be proud of. In 2019, it found the most massive neutron star to date, weighing 2.14 solar masses. In 2006, the GBT detected in space the compound acetamide, the largest interstellar molecule with a peptide bond, as well as other organic molecules, shedding light on our understanding of the chemical composition of the interstellar medium. The same year, the telescope discovered a large coil-shaped magnetic field in the Orion molecular cloud and a huge hydrogen superbubble—a cavity of hot gas that is less dense than the surrounding interstellar medium—in the constellation Ophiuchus, some twenty-three thousand light-years away.

Because of its excellent sensitivity, the enormous radio quiet zone all around, very low-noise receivers, and cutting-edge pulsar search technology, the GBT has been an amazing pulsar-finding and pulsar-timing machine. Over the years, GBT has discovered more than two hundred pulsars, many of them millisecond ones. It even found a number of millisecond pulsars in gamma-ray sources previously detected by Fermi, which means we have now identified some neutron stars that emit in both radio and gamma rays. Ryan Lynch, an astronomer based at GBT, calls it the best pulsar telescope in the world. So when it comes to finding a large population of millisecond pulsars in the Galactic Center, then GBT will be one of the instruments most likely to deliver—rivaled only by MeerKAT, and in the future the even larger SKA, and possibly China's FAST. But still, Slatyer says, detecting pulsars in the Galactic Center will be borderline even for them.[30]

In 2009, GBT did spot three pulsars not far from the center. They are not millisecond ones, but rather young neutron stars, and they are not gravitationally bound to the central black hole. Most researchers are more interested in millisecond pulsars specifically, because if they were bound to the black hole, it would allow them to perform precise tests of general relativity (see Chapter 8) by looking at tiny changes in the "ticking" of their clocks—or more precisely, the arrival time of their pulses.[31]

Still, the discovery of these three pulsars proved that GBT is (at least somewhat) able to peer through the mess shrouding the Galactic Center region. And messy it is—with huge amounts of gas and dust swirling around the Galactic Center, obscuring our view. More gas in the interstellar space means more free electrons zooming around, impacting the lower frequencies of the radio waves emitted by pulsars. That means the signal that astronomers receive will be greatly dispersed, smearing the pulse, and this smearing makes it harder to see the point source. Then there is even more interference by ionized gas, which scatters any signal. Pulses on their way to Earth are deflected by the gas, meaning they have to travel longer distances to get to us. They arrive at different times, which causes them to go from a really sharp looking feature to spread out in time. This effect makes it harder to detect the pulses—and if the scattering is severe, we won't be able to see any, because all of them will pile up on top of each other.[32]

The GBT, however, can deal with these effects more easily than many other telescopes. It's extremely sensitive to high frequencies, which are less prone to scattering and dispersion effects that would interfere with our ability to detect them, making it more likely it will find pulsars. On the downside, "the pulsars don't cooperate with us, because they get dimmer as you go to high frequencies," says Lynch. So there's this trade-off between trying to overcome these negative effects of the ionized gas between the Earth and the pulsars, but still trying to have enough sensitivity to detect the pulsars where they're already intrinsically faint. In other words, the game astronomers have to play is having a telescope that operates at frequencies both high enough to overcome the effects of ionized gas, and sensitive enough to pick up these fainter signals from pulsars, should they be there.[33]

Most millisecond pulsars are found in globular clusters, distant star formation regions in the Milky Way. Hooper says that based on how they shine, there simply can't be enough such pulsars in the Galactic Center to explain the excess. Others, however, do think it is possible. This debate was something of a stalemate—at least for a while, until 2015. Then two groups of scientists—one led by Christoph Weniger, an astrophysicist at the University of Amsterdam, and the other by Slatyer—presented pro-pulsar papers. Each group used a slightly different technique that worked within the technical constraints of the Fermi telescope. Because Fermi doesn't point at a source in a traditional sense but instead surveys a very wide area of the sky, it picks up radiation streaming from all directions. Particularly bright point sources (including pulsars) show up as hot spots in this radiation, but fainter pulsars can easily be lost in the background.[34]

Both teams split the region of the sky around the Galactic Center into many pixels. Then they measured the fluctuations in the radiation level from pixel to pixel, counting individual photons of light as they encountered the Fermi telescope. They noted that there was measurable variation between pixel-sized regions, even after factoring out variations in the expected emission from known sources: in other words, there were "hot" and "cold" areas of the sky. The hot areas, or bright pixels, they said, could be explained with the presence of one bright pulsar or a cluster of millisecond pulsars. Dim pixels—the colder patches—indicated no pulsars. While some hot and cold spots would be expected just by chance, the level of variability was high enough to suggest a population of pulsars just faint enough that they would be difficult to detect individually. "If you saw some blotchiness in the image, it was due to pulsars, even though you wouldn't be seeing individual pulsars efficiently," says Linden.[35]

Had dark matter been the source of all the excess radiation, the papers' authors argued, all pixels should have shown a uniformly smooth, hazy glow of gamma radiation coming from the still theoretical WIMP collisions. Pulsars seemed to have won the debate. Even Hooper had to admit that the pro-pulsar argument was pretty strong; he remembers that the two papers were very influential among scientists and convinced many that rapidly spinning neutron stars were probably the source of the radiation. While previously there used to be a new paper every few days about the dark matter interpretation on the arXiv, says Hooper, suddenly that rate went down by a factor of three or four. "Just like that, the bottom fell out on the interest in the subject."[36]

But Hooper continued to be skeptical and was very outspoken on what he saw as the limitations of both papers. "I thought a lot of things could explain this sort of clumping; after all, most of the gamma rays that this telescope detects are not from the excess that we're talking about, but from more mundane, astrophysical processes. And those probably have a lot of clumping associated with them that we don't know about." To him, it seemed totally plausible that hiding in the data could also be a lot of smooth excess from dark matter annihilation. He made the argument but had little success convincing others.[37]

Luckily for Hooper, Slatyer didn't put the debate to bed, despite her convincing 2015 paper. In early 2019, together with fellow MIT physicist Rebecca Leane, she decided to revisit her own and Weniger's previous calculations. This time though, she and Leane created a digital model of the Milky Way, complete with stars, gas, dust, and all the known pulsars. Then they injected the theoretical dark matter, along with some small pulsars that were not included in the initial digital mockup. They analyzed this made-up Milky Way—and found that the additional pulsars made it look like the galaxy contained very little dark matter shining in gamma rays, even though they knew it had to be there.

Then the researchers added the mocked-up dark matter data on top of the actual Fermi data, to see what would happen if this were applied to our real galaxy. And they again found a much smaller amount of dark matter than the signal they had added by hand—and far more pulsar-like point sources, indicating that a signal that should have been smooth appeared clumpy. They couldn't find any of the dark matter signal they had added until they injected more than five times as much as would be present if it explained the observed excess. So, they said, the paper shows that the previous analysis lacked the ability to find dark matter to begin with; its signal somehow remained hidden.[38] "You put it in by hand to be smooth, and it's saying it's clumpy. That just means that the analysis tool you're using is misidentifying smooth emission as clumpy emission," says Hooper. "It doesn't mean that it is necessarily smooth. But it certainly doesn't mean it's clumpy."[39] The analysis from 2015 turned out to be less robust than thought, and dark matter was still in with a chance.

While the paper presented no new evidence for dark matter, it weakened the pulsar-based explanation for the galaxy's excess gamma-ray glow. It's still unclear if the excess emission is clumpy or smooth—it's impossible to tell from the data available today, says Hooper, at least based on the data analyses done so far.[40] One future test could come from the data that Fermi is collecting while looking at the tiny galaxies orbiting the Milky Way, the dwarf spheroidals. If dark matter particles are responsible for the excess, then WIMPs in them should be making a very similar signal, just fainter. As Fermi collects more data and as we discover more and more of these dwarf galaxies, it might be possible to check it out.

For now, though, Slatyer's paper has reinvigorated interest in the dark matter interpretation. She herself thinks that pulsars are a more likely explanation—and Hooper acknowledges that many others are in that camp, too. "For every ten times that somebody finds a signal they say might be some new exotic discovery, nine of those times it will turn out to be something mundane. And that's just the reality of how this stuff works. Most of the time, you're not discovering new exotic physics but a new background you didn't understand," he says.[41]

Spotting dark matter would certainly be huge. It would open a whole new window onto a part of the Universe that previously we have been able to observe only by its gravitational effects; it would be an enormous advancement in cosmology. From seeing the gamma-ray glow, astronomers could estimate the mass of dark matter and estimate how it is linked to the Standard Model of particle physics, which best describes our understanding of how all known particles and three of the four known fundamental forces (the electromagnetic, weak, and strong interactions, but not gravity) are related to each other. "It would revolutionize my field," says Slatyer.[42]

Still, she doesn't agree with Hooper that pulsars are mundane—and thinks that finding a whole new population of millisecond pulsars in the Galactic Center would be a big deal. The detection would encourage astronomers to probe the Milky Way's evolution in order to figure out how these stars got there. "It feels really ungrateful to say, like, Oh, it's not dark matter, I didn't get the detection I wanted, so I'm going to turn up my nose," says Slatyer. "Finding a whole new population of neutron stars that we never knew existed would give us a lot of clues about the history of our galaxy. What I fear is that we may end up not knowing."[43]

So the race is on at full tilt. If over the next decade or so GBT, MeerKAT, Parkes, FAST, Arecibo, and future instruments such as SKA do discover pulsars all around the Galactic Center, then the score will be Neutron Stars 2, Dark Matter 0.

· 7 ·

When Pulsars Have Planets

ANDREW LYNE WAS NERVOUS. Sitting in a large conference room, brightly lit by huge chandeliers, he noticed that more and more people kept coming into the already packed room. It was a Wednesday, January 15, 1992, and at least a thousand people had come to listen to his talk at a meeting of the American Astronomical Society in Atlanta, Georgia. This was a special session of the conference, the third day devoted to the possibility of planets forming around neutron stars. Lyne, an astronomer at Manchester University, was on edge. Half a year earlier, he and his colleagues had announced that they had detected, for the first time ever, a planet orbiting a star outside our Solar System. The news was greeted as the first real evidence that there could be planetary systems other than ours in the Universe, even though this planet did not revolve around a star like our Sun but rather around the dead core of a once-massive star that had turned into a pulsar. When Lyne's team had announced their "discovery" in July 1991, the response had been ecstatic. And now Lyne had to get up and tell his audience that it had all been a mistake. He coughed slightly to clear his throat.

Their mistake had its roots in an event that occurred much earlier, in 1985, when the Lovell Telescope at Jodrell Bank detected a series of pulses from a new neutron star. Lyne and his colleague astronomer Trevor Clifton called it PSR B1829−10, put it in the catalogue, and as usual, started measuring the times of arrival of its pulses to determine the star's period of rotation. These times changed as the Earth moved around the Sun. Similar to timing a typical binary system, when our planet was on one side of the Sun, closer to the pulsar, the pulses arrived earlier at the telescope; as the Earth moved to the other side, the pulses arrived a little bit later. By timing the pulses for several months, they managed to pinpoint the pulsar's rough position in the sky. It lived in the constellation Scutum, about thirty-five thousand light-years from Earth.[1]

In May 1991, however, the astronomers noticed something strange about the pulsar's period: every 180 days or so it fluctuated. At first, they dismissed it as timing noise, the rotational irregularities seen in most pulsars; they can happen randomly, but astronomers don't quite know why. Then a graduate student, Setnam Shemar, after studying the data, suggested that the oscillations might be caused by a body with the mass of a planet that was orbiting the neutron star every half a year. Lyne and Matthew Bailes had observed the pulsar for several years and the odd had period stayed the same, that is, it was stable in its amplitude. "The only explanation we could think of was that it was due to the motion of the neutron star in orbit with a very small body with a planetary mass," Lyne tells me as we sit in Jodrell Bank Observatory in July 2019, with the Lovell Telescope towering above. Lyne looks tired. It's clear that it's painful for him to recall the events.[2]

The planet, Shemar calculated, would have the mass of Uranus and an orbit similar to that of Venus—going around the pulsar every half a year. A planet orbiting a pulsar. Mind-blowing.

Lyne, Shemar, and their other collaborator, Matthew Bailes, published their paper in *Nature* on July 25, 1991, with the journal cover line "First Planet Outside Our Solar System." The discovery triggered a media frenzy, with journalists from mainstream and science outlets alike reporting the find for weeks, and other researchers hailing the work.

Still, something about the six-month period just didn't seem right; it kept nagging at Lyne. He worried that somehow he and his colleagues had made a mistake and that the odd changes in the pulse might simply be the result of our Earth's motion around the Sun. Over the following Christmas and New Year's holiday, when he finally had some quiet time, Lyne decided to revisit the calculations.

On January 2, 1992, he re-analyzed the data in more detail, and noticed that the astronomers had worked with two slightly different positions for the pulsar, differing by one ninth of a degree. Usually, the two positions are updated when a new position is found, but not in this case. And the omission affected the whole timing model, since the precise time when a pulse gets to Earth depends on the relation between the Earth's position in its orbit around the Sun and the neutron star's exact location in space.

When Lyne entered the pulsar's new position, suddenly the six-month periodicity disappeared. He and his team had simply misinterpreted the cyclic variations of the pulses, having explained it with the gravitational effects of a companion planet. They had failed to correctly compensate for slight variations in Earth's orbital motion around the Sun. There was no planet around that pulsar.

"I just—I just couldn't believe it," says Lyne. "I was there all by myself up in the room along the other end of the observatory. And you can imagine all the sorts of things that went through my mind after all the hullabaloo and everything else. And then I realized our discovery wasn't the truth."[3]

The mistake was due to the fact that Earth's annual orbit isn't perfectly circular. It's slightly elliptical, off by about 1 percent. But if you approximate it to being circular, you need to make a correction on the order of seconds. There was an error in the calculation of this correction on the order of milliseconds and this had the six-month periodicity. When Lyne moved the pulsar's reference position to its actual position and repeated the analysis with the correct motion of the Earth around the Sun, it became clear that there was no "planet" there. It was a stand-alone neutron star.

As Bailes remembers it, a few weeks before the 1992 conference, very early in the morning, Lyne showed up at his house in Cheshire. "You know why I'm here," he told Bailes. "The planet?" Bailes asked, nervously. "Is it . . . gone?" Lyne sighed. "I'm afraid so."[4]

Lyne and his team had to retract their paper, and the opportunity came at the conference, where Lyne had been invited to speak about the discovery of the new planet. The reversal was not what the audience was expecting. Even before he finished speaking, journalists had started running out of the room to file their stories: the first extrasolar planet had not been detected after all. The stunned audience applauded. "People thought I was brave, but I had no choice. I knew that other people would be looking at this and very soon they would come to the same conclusion that I did. Someone else would have found it soon," Lyne sighs. For Bailes, and probably also for Lyne, "recounting the experience is still like a stake through the heart," says Bailes.[5]

As Lyne stepped off the stage, he glanced at the next presenter. It was Alex Wolszczan, an astronomer at Cornell. Lyne, Wolszczan, and the president of the astronomical society, astrophysicist John Bahcall who was the event's organizer, were among the few people in the room who had known in advance about Lyne's retraction.[6]

Wolszczan walked on stage and fired up his images. Until the day before, he had felt like a runner who was arriving second at a marathon's finish line—except late that evening, only hours before Lyne and he were to present, he found out that the marathon's winner had just been disqualified. Wolszczan cleared his throat and told the audience that it was in fact he, Alex Wolszczan, who had detected the first extrasolar planet—and not just one, but two (later, he would find a third one). These planets, he said, were orbiting a millisecond pulsar. And this time, the detection was very real.

I AM 150 METERS (NEARLY 500 FT) ABOVE THE GROUND, on a narrow suspension bridge that is swaying in strong winds. The metal mesh

Alex Wolszczan in front of the Arecibo Observatory. (Courtesy of Tony Acevedo, Arecibo Observatory)

railing on both sides is mildly reassuring, but the height makes me gasp. Better not let anything drop, I think. There would be no way to retrieve it—and it could damage the highly sensitive dish below, which is gigantic at 305 meters (1,001 ft) across. Welcome to Arecibo Telescope in Puerto Rico's rainforest—the telescope that Alex Wolszczan used when, in 1990, he found extrasolar planets orbiting a pulsar. His detection changed our understanding of planet formation.

The bridge I'm crossing carefully is the one that Wolszczan used to sprint across at least twice a week between 1983 and 1992, to reach the platform suspended high above the center of the dish. He was one of Arecibo's resident astronomers back then, and going to the platform was essential for checking the state of the receivers and other equipment. "I did it for all kinds of reasons, but it was always a pretty thrilling experience to be up there, because it's such a gigantic construct," he remembers.

"And you really get a different perspective looking at it from almost above."[7]

Stepping onto the platform makes me feel slightly lightheaded and a little bit like James Bond, because it's right here on this bridge and the platform that 007 battled it out with the master villain in the movie *GoldenEye*. Except that in real life, Bond actor Pierce Brosnan, due to a fear of heights, never walked on this bridge—his stunt double did. I grin when I'm told that I achieved something that James Bond couldn't do. Another disappointment for Bond fans: the dish never emerged from an artificial lake—that was done with a miniature model of Arecibo. Engineering manager Luis Quintero fills me in on a few other inaccuracies. The movie imagines the dish as being in Cuba, not Puerto Rico; and in real life, no one has ever run across the dish itself, let alone fallen onto it. One big difference is not the movie makers' fault: the triangular platform now holds up steerable receivers housed in a beautiful Gregorian dome that was installed in 1997, two years after the movie opened in cinemas.

While it was movie trickery that damaged the telescope in *GoldenEye*, Arecibo did receive a severe beating in September 2017 during a natural disaster. When the deadly Category 5 hurricane Maria devastated communities in Puerto Rico, Dominica, and the US Virgin Islands, it didn't spare the observatory, causing some $14 million in damage. A huge chunk of a pointy 430 MHz feed line used for atmospheric studies was torn off and crashed right through the dish; some ninety reflector panels had to be replaced. The space under the reflector also flooded, which damaged a lot of the cabling and heating facilities. It's in the process of being fixed, Quintero tells me, but it's taking a long time.

The dish, or to use its technical name, the fixed spherical reflector, has been built into a natural depression in the ground—the result of a karst sinkhole. Completed in 1963, the observatory was originally designed for national defense during the Cold War—the US government wanted to use it to detect Soviet satellites and missiles. In 1974, the telescope became world famous when it transmitted the Arecibo message—the most powerful broadcast ever sent into space—devised to be picked up by aliens in the neighboring globular cluster M13. Sadly, there are still another 24,950 years or so to go until the message arrives at its destination, by which time the stars in M13 will have moved on from that location.

The surface of the dish is covered by 38,778 aluminum panels, each one by two meters (three by six feet). Under the reflector I find a dense tropical rainforest, with giant bamboo, three stories high; lizards the size of a hand; and grasshoppers as big as a finger. Walking underneath the reflector I can see how the surface is held together. It's a weird feeling, being underneath. It's like standing in a huge covered stadium, except the roof has been placed upside down. It makes you appreciate "how really a fragile thing it is, and huge at the same time. And then you realize what a gigantic job it is to keep this surface very close to the spherical shape, just by regulating the tension of those wires that hold the aluminum panels which comprise the dish itself," says Wolszczan, who spent many an hour reflecting under the mega reflector. Because the dish itself isn't steerable, scientists instead move the small dome above it up and down the banana-like azimuth arm, which can be turned. That way astronomers track sources in the sky, even though the gigantic reflector is fixed.[8]

Dotted around the telescope in this remote, radio-quiet zone in the rainforest, about two hours' drive from San Juan, are a few small buildings. To reach them, you have to follow a road that twists between cliffs. I come to the control room, a one-story building with a terrace that opens to a stunning view of the dish itself and the suspended Gregorian dome. It's in this control room that residents and visiting astronomers used to spend hours and days at their computer screens, slowly and precisely moving the dome into the position they needed for their observations. These days, hardly any scientists show their faces. As with most other telescopes, observations are done remotely, with the support of telescope operators on site. On a board, I see the names of many of the scientists whom I've interviewed for this book—all using Arecibo, but from the comfort of their offices around the world.

Still, there is a tiny canteen, where the cook graciously agrees to make me a vegetarian burger. A swimming pool is a nice and important touch, because the sun in this part of the world is scorching and unforgiving. And there still are a few houses where astronomers and visitors stay should they come here after all. Wolszczan was living in one of them—small, set on poles, and hidden in the jungle—when one morning in 1996 there was a knock on his door. It was the production crew of *Contact,* starring Jodie Foster as an astronomer who tries to detect alien signals here at Arecibo. Filming hadn't started yet, but Wolszczan's house had been selected as the location where Foster's character would live at Arecibo. "They asked me if they could actually go into the little house that I occupied, because Jodie Foster would stay there," he recalls, with a chuckle. "Then later I saw how they remade that little house inside—they made it into a little luxury palace out of something really very simple. It was pretty entertaining—kind of funny and amusing to see what they did for the movie."[9]

THE LIST OF SCIENTIFIC ACHIEVEMENTS associated with Arecibo is long. It was here that researchers detected a pulsar in the Crab Nebula in 1968, just months after the Vela Pulsar was spotted in the Vela Nebula—with both detections providing overwhelming evidence that neutron stars are very real and born in supernovas. It was Arecibo that in 1974 spotted the first binary pulsar, proving indirectly the existence of gravitational waves and thus confirming Einstein's theory of general relativity—a find that earned the researchers, Russell Hulse and Joe Taylor, the Nobel Prize in Physics. That very same year, Arecibo sent its SETI message in M13's direction. And in 1982 it was also this telescope that spotted the very first millisecond pulsar, which is spinning around itself a whopping 642 times per second.

And it was with Arecibo that Alex Wolszczan found the first planet orbiting a pulsar.[10]

It was June 1990, and Wolszczan had been busy running a survey in search for millisecond pulsars above and below the Galactic Plane. One of the aims was to test his theory that millisecond pulsars, because they are old, would have enough time to migrate out from their birthplace in the Galactic Plane into the space above and below it. Those places, he reckoned, should be full of them. At the time, only four millisecond pulsars were known. Wolszczan wanted to find more. "It was really a completely new idea to look away from the plane and try to find new pulsars and actually confirm the expectation that there should be a lot of them at high galactic latitudes," says Wolszczan.[11]

To prove that, he needed to use a telescope for a substantial period of time: a tricky constraint. Telescope time is hard to get. Astronomers have to submit proposals explaining their research, and the winning proposals typically net scientists only a few hours or days of observation time. But Wolszczan was lucky—he was able to run his survey for almost a whole month because Arecibo needed to undergo repairs. The supports holding the triangular platform above the reflector had developed cracks due to metal fatigue and needed to be replaced, so the telescope was not open to outside researchers. The few resident astronomers had the instrument all to themselves. While it wasn't possible to move the receivers to track sources by pointing at specific sections of the sky, the dish could still observe the sky as it was drifting by. And because the reflector is huge, it has very good sensitivity—"even though any point in the sky in that particular configuration was visible for only about thirty seconds," says Wolszczan.[12]

During his brief survey, he came across two millisecond pulsars, very high above the Galactic Plane. One of them piqued his interest more than the other: it was a binary system, only the fourth one found at the time, and certain characteristics made it a great system for testing Einstein's theory of general relativity. "I was really very, very occupied with that," Wolszczan says. "And I set the other pulsar that I had discovered aside for a little while, not knowing what was in store for me with it."[13]

As it turned out, this other briefly disregarded pulsar—which was only the fifth millisecond pulsar discovered at the time—would change not only Wolszczan's life, but also our understanding of planets and galaxies.

When Wolszczan finally found time to return to the other pulsar's data set, he noticed certain peculiarities. Just as with any new pulsar, he started to time it over several months, measuring pulse arrival times to construct a model. "But that particular pulsar clock behaved in a way that was very difficult to model," says Wolszczan. "It just didn't behave the way I expected, or anyone at the time would expect." The pulses were irregular, which was odd since millisecond pulsars are typically very stable. Wolszczan also noticed what seemed like a pattern in these irregularities, but he couldn't quite grasp what it was.[14]

Because millisecond pulsars are typically found in binary systems, astronomers know that the pulse arrival times differ from a pulse if there is no companion. Wolszczan was seeing such deviations all right, but they were very small. The less massive an object orbiting another star is, the smaller those deviations are, and the deviations he was recording were way too small to be caused by a white dwarf star, the usual binary friend of a millisecond pulsar. That was not the only issue. When Wolszczan tried to fit a single orbit to the data, it would work at first, but then the data points would no longer fit. He tried to improve the model, again and again, but the same thing kept happening.

Perplexed, he started observing the pulsar very closely, every day over the course of three weeks, to determine the exact shape of the deviations. He realized from this painstaking work that what he was observing was definitely not a standard binary pulsar. He also dismissed as very unlikely (since it was an old millisecond pulsar) another common assumption at the time: that the seismology of a neutron star could cause it to behave weirdly, producing so-called timing noise. Wolszczan next looked for possible mistakes in his data analysis and any technical problems with Arecibo's setup—but the other pulsar binary that he had discovered did not suffer from any of these odd effects. If neither the instrumentation nor the data analysis was at fault, the pulsar system had to be different somehow.

Despite Arecibo's sensitivity, Wolszczan couldn't pinpoint the exact location of the pulsar—but doing so was important, because any position error could be misinterpreted as an orbiting object. To get to the bottom of this, he asked Dale Frail, a fellow astronomer working with the twenty-seven dishes of the Very Large Array in Socorro, New Mexico, to give him a hand. It didn't take Frail very long to get a fix on the location.

Armed with the exact coordinates, Wolszczan had to take a scientific leap. "I saw very clearly that what I was looking at was something like a sum of two sine waves, beating against each other. So I knew that it must be two objects orbiting the pulsar," he says. When he measured the amplitude of those waves, he realized that they had to be of planetary—Earth-like—mass.

When he finally had the results, Wolszczan stared at his screen in disbelief. At the time, astronomers didn't really expect that planets could form around neutron stars, not least because Andrew Lyne hadn't yet published his (erroneous) paper; that would happen later that year. So while Wolszczan thought that two bodies of planetary mass could explain the data he was getting, he didn't believe that this could be the correct solution. He pondered the meaning of his data while on the platform at sunset, looking down on the dish; at one of Puerto Rico's famous sandy beaches—his typical weekend retreat; and when he stared at his computer screen, running various alternative scenarios but discarding them one after the other. He wanted to pre-empt any possible questions and criticism from the rest of the scientific community that were sure to come once he announced such an extraordinary explanation for the pulsar's behavior.

He left Puerto Rico in September 1991, going back to Cornell in Ithaca, New York. While there he crunched a full year's worth of data gathered starting in June 1990, testing his timing model under the assumption that he was looking at two Earth-mass planets in circular orbit. He ran one more complete and decisive test, and finally it finally dawned on him: no alternative explanation worked. The only thing that made sense was the hunch he had had months earlier—that he had discovered a dead star that was being orbited by two planets. It was, he says, his very personal eureka moment, right there in his office at Cornell. He had made a momentous discovery: a planetary system outside our own Solar System. "I looked at the result and it was absolutely perfect," he says. "I knew that I finally had it. I had to face the reality that [it] was just that—it had to be planets. Period." The masses of the planets were initially measured at 3.4 and 2.8 times Earth's mass.

He knew that two months earlier Lyne had published his discovery of a planet around a pulsar, so his find, Wolszczan thought, wouldn't be a first. He was okay with that. He and Frail published their paper on January 9, 1992, just days before Wolszczan's AAS talk.[15]

As Wolszczan traveled to the conference, he knew that he was scheduled to speak after Lyne. Not a confident public speaker at the time, Wolszczan practiced his talk again and again in front of a mirror. He had arrived in Atlanta a day before the event, but just as he was about to get himself a drink, the president of the astronomical society, John Bahcall, pulled him aside for a private chat. He asked Wolszczan to sit down. "He told me that Andrew would be coming, but not to confirm his discovery—to retract it. And he was kind of trying to impress on me that I should really be nice and accommodating in a situation like that."

The news about Lyne's upcoming retraction gave Wolszczan a few precious hours to modify his talk. That same evening, he approached Lyne for an awkward but important chat. "He was just sorry that it happened the way it did," says Wolszczan, "and that's completely understandable. He must've been upset by all this, but he didn't show it." There was no visible drama.

When Lyne finally stood up and announced his mistake, he even referred to Wolszczan's discovery, saying that this system would probably withstand any criticism or scrutiny. By then, though, his audience was already getting off their chairs. "It was really quite an uproar," recalls Wolszczan. When he himself took to the stage, Wolszczan was worried that he would face a lot of tough questions, but he didn't. Maybe, he wonders, it was because people were still so stunned by what Lyne had said, or maybe because he had meticulously gone through all possible alternative explanations and demonstrated why they didn't work. His scientific peers were satisfied with his planetary conclusion.

Wolszczan's discovery was more than finding a curious case of planets spinning around a leftover core of a massive star. It also proved that planets can exist around many different types of stars, forming just like they are born around stars like our Sun—from the disk of matter surrounding the star. The difference, though, is how this disk is created. Currently, scientists have two leading theories: either there's a so-called fallback disk, which may form out of some material ejected during a supernova explosion that fell back toward the dead core instead of flying into space, or a newborn neutron star can collide with the parent star's companion—another star. In that case, the star would be disrupted and start splashing its own material around, forming the disk.

Researchers who produced papers based on Wolszczan's findings soon said that planets can probably form around any kind of star imaginable; the process of planet formation must be universal. Years later, the Kepler space telescope would confirm this, spotting plenty of planets orbiting all kinds of stars, including white dwarfs. Wolszczan's own system, to which he later added a third planet that he hadn't detected at first, was sort of a preview of what Kepler would find two decades later: the most common planetary systems are actually super-Earth systems (with masses just above that of our Earth) in very tight orbits around their star. In the 1990s, nobody had expected that. "Back then, if you said, 'Oh, in twenty-five years or so you will see that planetary systems would mostly look like that,' nobody would have believed it," laughs Wolszczan.

Could there ever be life on planets like Wolszczan's? He finds the idea far-fetched, but in the past few years there have been papers exploring the possibility of habitable zones around neutron stars. Perhaps one day we may find life spinning around a pulsar—maybe not a full-blown civilization, but at least some microbial, primitive life. In that case, the interstellar medium would reveal to us yet another mystery of our peculiar and mesmerizing cosmos.

8

Giant Scientific Tools of the Universe

NEUTRON STARS MAY BE FASCINATING TO STUDY, but for some scientists, there's much more to them: they want to use them as tools on a galactic scale.

One of the potential applications is to use pulsars as cosmic beacons for a global (or should that be galactic?) positioning system. It's been shown to work in principle and could support future space navigation to Mars—the dream of eccentric billionaire entrepreneur Elon Musk and others—and beyond.

Here on Earth, many car drivers, truckers, pilots, ship captains, and most owners of smartphones find their way around by using navigation systems like GPS, Galileo, and GLONASS. These systems work by timing the signals sent out from ultra-precise atomic clocks that orbit the Earth on satellites. Taking the concept into space, some researchers have long been proposing to use the ultra-precise "ticking" of the most accurate pulsars to help spaceships find their way in the vastness of the Milky Way. Right now, a spacecraft has to communicate continuously with Earth, via the giant satellite antennas of NASA's Deep Space Network, to make sure it is still on the right track. The farther the craft ventures into space, the trickier, less reliable, and more costly this technique becomes. By using pulsars as beacons, a probe could position itself autonomously. It would be the ultimate space compass.

In November 2017, NASA showed that using pulsars as beacons is possible—with the help of NICER, an x-ray detector aboard the International Space Station. The detector is busy most of the time determining the mass and radius of neutron stars, in a bid to help scientists understand the dead stars' internal structure. But NICER also sports a little tool called the Station Explorer for X-ray Timing and Navigation Technology (SEXTANT) that can very precisely monitor pulses from multiple neutron stars scattered across the sky. That's what it did in 2017, timing the x-ray radiation from five pulsars, each for five to fifteen minutes. SEXTANT measured minuscule changes in the arrival time of the signals and as the ISS orbited the Earth, the instrument calculated NICER's—hence its own—precise location. In the future, a similar, lightweight tool could be easily added to any spacecraft.[1]

In addition to aiding interstellar travel, timing pulsar arrays could also help shed light on a mystery that astronomers have been trying to crack for decades. Researchers know that in the center of most galaxies lurk supermassive black holes millions or even billions of times the mass of the Sun. The black hole at the center of the Milky Way, for example, is roughly four million solar masses.

As galaxies collide, astronomers think these black holes will merge—and as they do, send out gravitational waves across the cosmos. LIGO and Virgo on Earth won't be able to spot these waves, however, because the waves will be much too long and too weak to ever be picked up by a ground-based detector. So since the 1980s, astronomers have been using pulsar arrays to spot such ripples in spacetime, effectively turning neutron stars into galactic-scale natural gravitational wave detectors—predominantly sensitive to black hole systems that are in the size range of 100 million to 10 billion solar masses.

If they ever do detect such a wave, scientists would be able to test Albert Einstein's theory of general relativity on an unimaginable scale—and confirm the existence of supermassive black hole binaries. Such an observation could also tell us more about the enigmatic, crowded Galactic Center environment they live in. Pulsar timing arrays might even help us unlock more secrets about the objects that could be responsible, if only partly, for the very formation of supermassive black holes—neutron stars. After all, when a massive star dies, it either collapses into a stellar-mass black hole straight away or forms a neutron star, and when two neutron stars collide, a black hole is (typically) born. One theory is that many mergers of such stellar-mass black holes eventually leads to a supermassive monster.[2]

To try to understand these galactic beasts, let's put pulsar timing aside for a minute and take a trip to the edge of the world.

THE ANTARCTIC SNOW IS BLINDING, and a few hundred yards in the distance, the ten-meter radio dish looks tiny and so white that it nearly blends into the barren, frozen plateau. This is the South Pole Telescope, and its job is to spot coming in from space really short waves, those ranging from the millimeter to submillimeter part of the electromagnetic spectrum. While the telescope was originally built to measure the faint signatures of the oldest light in the Universe that remained after the Big Bang, the cosmic microwave background radiation, lately it's been part of another adventure. Linked together with a bunch of other antennas positioned all around the globe, the South Pole Telescope is part of a virtual radio dish about the size of the Earth, the Event Horizon Telescope (EHT). It's this giant virtual tool that has allowed scientists for the first time to get up close and personal with a colossal supermassive black hole at the center of a neighboring galaxy—and snap its picture.[3]

Just three months before the image became public on April 10, 2019, Dan Marrone, an experimental astrophysicist at the University of Arizona, was trying to make his way to the South Pole Telescope without getting lost. It was a windy morning in late January; Marrone had just walked out of a two-story building, the US Amundsen-Scott South Pole station in Antarctica, home to about 150 people who live and work there for many weeks at a time. Luckily, in January the sun is up nearly all the time, because it's summer in the Southern Hemisphere. Still, with temperatures at nearly −20° F (−29 C), the wind had sent heaps of snow swirling all across the Antarctic Plateau. The astrophysicist was walking slowly, with ten thousand feet of ice below him, across a flat plane of white nothingness. In order not to lose his way—which is dangerous in summer and deadly in winter—he followed the flags planted along the entire route to the dish. "In the winter, it gets much more dicey here," he says, now back in Tucson, Arizona. "It can be minus 100° F [−73° C], and the visibility can go to almost nothing without any light. I imagine that can be quite scary. But fortunately, in the summer, I never have to deal with that."[4]

He finally made it to the telescope, and a few hours of work later, with the wind much calmer, he walked back to the station about one kilometer (nearly two-thirds of a mile) away. The station is its own little world, complete with a gym, language classes, a ping-pong table, and satellite-based internet (that only works for a few hours a day) to stay connected to civilization. Snow never thaws there; every year, another twenty centimeters (nearly eight inches) accumulates, settling under its own weight. That's why the 7,400 square meter (80,000 square ft) building is elevated, propped up by several support columns that ensure it won't get buried. Traveling here is far from trivial: Marrone first flew from the United States to Christchurch, New Zealand, on a commercial flight, then hopped onto a small ski-equipped US Air Force LC-130 Hercules propeller plane to McMurdo Station on Ross Island, and from there—after a wait of a night or more, depending on the weather—onward with the LC-130 to the South Pole.

It was Marrone's fifth visit to Antarctica, but he hadn't been coming to the bottom of the world just for the thrill of it. While the SPT has been around since 2007, it was Marrone who in 2010 had the idea to link it virtually to several other ground-based radio dishes as part of a planet-scale array, the EHT. The term "event horizon" describes the virtual boundary of a black hole—the spheroidal region of space thought to be left behind after a massive star (more massive than those that give birth to neutron stars) has been crushed by its own gravity. In a black hole, all mass is concentrated in a very small region at the center called the singularity (in the case of a spinning black hole, this singularity is a ring). The black hole's boundary—the event horizon—is the critical edge that, once crossed, won't allow anything, not even light, to escape.[5]

On April 10, 2019, the EHT collaboration made history when it published what at first glance looked like a reddish-orange doughnut on black background. While seemingly unimpressive and blurry, the picture was the first-ever high-resolution image of the shadow of a black hole—the supermassive one at the heart of the biggest galaxy in the nearby Virgo galaxy cluster, Messier 87 (M87). (I get the news while connected to the internet in the Bunker, the isolated room near MeerKAT in South Africa, which for me personally makes this milestone even more significant.) The black hole is 55 million light-years away from us and 6.5 billion times more massive than the Sun. Scientists have long thought, based on Einstein's general theory of relativity, that a black hole should have a dark shadow-like area that is immersed in a bright region of glowing gas. Now they have their first proof.[6]

Gravity bends the light, and the Doppler effect increases the intensity of that light for material rotating toward the observer—us. The light that might come to us from behind the black hole can't pass through it, so much of it instead gets lensed (or bent) around it. This means that very little emission can appear directly in front of the black hole, because any direct emission would be bent away from our line of sight. "The light from in front of the black hole is mostly redirected around it by gravity or captured by the event horizon," says Marrone.[7]

Because the black hole is so far away, scientists needed a huge, ultrasensitive instrument with very high resolution to image it. The EHT is just that. It relies on very-long-baseline interferometry, which it achieves by synchronizing multiple telescopes around the globe. Making use of our planet's rotation, the array becomes an Earth-size instrument with an angular resolution of 20 microarcseconds, able to detect wavelengths as tiny as 1.3 millimeters.[8]

Marrone thought that the South Pole Telescope would give the EHT a major boost, especially for observing the Galactic Center of the Milky Way and its supermassive black hole, Sagittarius A* (Sgr A*). The astrophysicist has a special relationship with Sgr A*—he wrote his PhD thesis on it. In November 2011, he submitted a proposal to the National Science Foundation that runs the SPT, asking for a grant. He explained that the telescope would be key to snapping a picture of our own black hole, because it would double the resolution of the array. After all, our Galactic Center is best seen from the Southern Hemisphere, and the South Pole is as far south as one can get. The tiny Antarctica-based telescope can observe the Galactic Center all the time.[9]

Researchers have long known about Sgr A* and its home, the center of the Milky Way. In 1931, Karl Jansky picked up a radio signal from that direction. Because it was suspiciously close to the constellation of Sagittarius, the radio source was dubbed Sagittarius A—which then morphed into Sgr A* once astronomers identified a point-like component within SgrA. Over the years, plenty of researchers theorized that the concentrated component was tell-tale sign of a black hole, generated as it consumes the matter around it. Several teams published papers in the last few decades tightening the radius limits on the central mass, both theoretical and observational—based on the measurements of the velocities of stars near the center of the Milky Way.[10] Then, in 2002, a team led by Reinhard Genzel at the Max Planck Institute for Extraterrestrial Physics published a paper based on a decade-long study of the motion of stars around the Galactic Center, especially a star called S2. Their data suggested that S2 was orbiting a very compact and bright central radio source, some 60 million km (30 million miles) in diameter, that was too compact to be a very dense clump of stars. This was yet another piece of indirect evidence that the source is, in all likelihood, the extreme environment surrounding a supermassive black hole, some 26,000 light-years from Earth.[11]

It took half a year for Marrone to get his 2011 application for a grant approved—after which he got busy developing and building all the necessary equipment for the little dish to spy on black holes. Specifically, he designed a coherent receiver to detect the electric field of whatever source the telescope points at. The receiver is like "a very fancy radio that's at four degrees Kelvin," says Marrone—meaning that it's nearly as cold as interstellar space, to minimize any noise.[12] It's this receiver combined with an ultra-precise atomic clock that Marrone works with every time he operates the telescope.

While M87 is not visible from Antarctica, the South Pole Telescope's analysis of a different source in the sky, the quasar 3c279, greatly boosted astronomers' confidence in the final result. "Otherwise, I think we would have been significantly more nervous," says Marrone. In the six papers that have come out around the same time as the doughnut-shaped image, the researchers have revealed the mass of M87's central black hole and the direction of its rotation. They also now have, says Marrone, the first glimpses of its immediate environment, allowing astronomers to better understand how black hole jets work.[13]

During the years of eyeing the center of M87, the eight telescopes of the EHT have also been zooming in on Sgr A*. Combined with the data from ALMA on top of the Chajnantor Plateau in Chile, the results should soon allow the researchers to create an image of our own black hole—or rather, its shadow. With these images, researchers are hoping to learn more about the environment and the workings of supermassive black holes—the ultra-hot gas and dust they bathe in and then gobble up, as well as the huge jets they spit out when the incoming gas gets accelerated.[14]

How supermassive black holes form is a mystery. But even much smaller, stellar-mass black holes, which we know are born out of massive stars that have collapsed and died, were at first merely mathematical concepts and curiosities. In 1916, the mathematician Karl Schwarzschild solved Einstein's equations of general relativity for a spherical mass (the now famous Schwarzschild radius of a black hole). And in 1958, a physicist at the Georgia Institute of Technology, David Finkelstein, showed that black holes had an imaginary boundary beyond which not even light can escape—the event horizon, which permanently separated stuff that has fallen into a black hole from the outside Universe.

Still, for decades black holes—both stellar-mass and supermassive—were merely theoretical (albeit widely accepted) concepts. Only in the middle of the twentieth century did astronomers start gathering evidence of their existence, such as the first observations of quasars. The galactic x-ray source Cygnus X-1, discovered in 1964, was later determined to be most likely a black hole. LIGO's detection of gravitational waves from the collision of two stellar-mass black holes on September 14, 2015, really helped, too. Now, with EHT's picture, we know that supermassive black holes are also truly there.

But how do they get so big? Scientists don't know, though one suggestion is that they stem from many stellar-mass black holes of tens or hundreds of solar masses that may have been merging in the early Universe. As they merged, accreting gas in the process, they kept on growing—leading eventually to a supermassive black hole.[15]

Throughout the evolution of the Universe, galaxies are also thought to have been colliding and merging. Our own Milky Way, for example, is set on a collision course with the neighboring Andromeda galaxy; the two will become one in about 4.5 billion years. Much farther out from Earth, about one billion light-years away, astronomers have even spotted a rare system of three galaxies known as SDSS J0849+1114—each with its own supermassive black hole. The three seem to be set for an almighty collision.[16]

When galaxies merge like that, their central black holes should merge too—morphing into an even more supermassive behemoth (especially if there is a third black hole nearby that can tug them closer together). Based on computer simulations, really massive black holes may have undergone as many as twenty mergers during their lifetimes. M87 may have had a few mergers too, since it is billions of times more massive than the Sun—or any stellar-mass black hole. At the same time, astronomers believe that a supermassive black hole will acquire most of its mass over time by ingesting interstellar gas that falls toward it when its host galaxy is disturbed by another. "We're not exactly sure what the ratio between those two is—accretion of gas and the mergers," says Scott Ransom, an astronomer at the National Radio Astronomy Observatory and the University of Virginia.[17]

That's all theory so far, but the thinking goes that when galaxies merge, the in-spiral and eventual collision of their supermassive black holes should trigger gravitational waves in all directions. We know that massive bodies in space disturb the fabric of spacetime, creating a well. A single mass on its own doesn't produce gravitational radiation, but the motion of two accelerating masses around each other can. The larger the disturbance, the more spacetime will react, sending out more energetic waves.

As I described earlier, Hulse and Taylor showed the existence of gravitational waves indirectly, when they noticed the shrinking orbit of the binary pulsar they had detected; LIGO then caught gravitational waves directly from the merger of two stellar-mass black holes in September 2015. And Marica Branchesi found out about gravitational waves coming from the first-ever observed collision of two neutron stars while at hospital in Urbino, the night of August 17, 2017—while her sister was giving birth.

The gravitational ripples from a hypothetical merger of two supermassive black holes, however, would be much, much longer and change more slowly, with frequencies in the nanohertz range. The detector arms of LIGO and Virgo would have no chance of detecting them. When stellar-mass black holes merge, they are spiraling in on each other, revolving around their common center of mass many times each second. The gravitational waves they give off during the in-spiral and the collision are very high in frequency; they range from 7 kHz (a wavelength 43 km, or 27 miles, long) to 30 Hz (a wavelength 10,000 km, or 6,214 miles, long). Our ground-based detectors then measure how much space is squeezed over the length of each of their arms. For waves longer than the arms, LIGO measures the squeezing over much less than one wavelength, so the detectors are measuring the stretching between two points close together on the wave, essentially measuring the "steepness" of the wave coming in. As the waves stretch, the longer wavelength means they are less steep and LIGO is less sensitive to them.

When a supermassive black hole of a billion solar masses finds itself in a binary system with another black hole of a similar size, however, they would be orbiting each other for years before crashing into each other. Their in-spiral and merger would emit gravitational waves that are light-years, hundreds of trillions of kilometers, long and of extremely low frequency—there is no way LIGO or Virgo or any other ground-based detector would notice if one were to wash over it.[18] The main issue is noise from Earth, which easily masks waves of lower frequencies. Noise can be anything from minor earthquakes all over the globe, to ocean tides and even trucks driving by. "A couple times per second is basically the limit as to the lowest frequencies that LIGO can see—because below that, you're just dominated by the noise of the Earth and seismic activity," says Ransom. "It just overwhelms everything else."[19] In space, though, one can optimize the system to be stable on these much longer timescales.

And that's where pulsars come to our rescue. While neutron stars are themselves sources of gravitational waves when they spiral toward each other and then collide, they can also be used as galaxy-sized detectors of gravitational waves—at least that's the hope. Researchers are optimistic that over the next decade, using the technique of pulsar timing, they will finally spot gravitational waves from a merger of two supermassive black holes.[20]

The Solar System as a Timing Array

In 1969 and 1970, Joe Weber—an electrical-engineer-turned-physicist—announced to the world that his barrel-shaped aluminum cylinders had detected gravitational waves. Unfortunately for him, he was wrong.

Back then, the first discussions about building LIGO were still another decade away. Two astrophysicists, however, were pondering whether they could use the cosmos itself to spot spacetime ripples. One was Soviet researcher Mikhail Sazhin; the other, US scientist Steven Detweiler.

Within a few months of each other, in 1978 and 1979, they published papers suggesting that timing the arrival of pulses from neutron stars might help astronomers observe gravitational waves. For that to work, they said, one would have to think of the Solar System as the center of a scientific instrument (like the intersection of LIGO's two arms) with a faraway pulsar at the end of one virtual arm stretching across interstellar space. The pulsar, being an extremely accurate clock, sends out regular pulses—and if ever there is a gravitational wave passing through, local spacetime would be disturbed, and the pulses would then arrive early or late compared to what astronomers would have otherwise predicted.

Two other theorists, Ronald Hellings and Gabriel Downs, later refined the idea, extending it to monitoring an array, or a set of pulsars. Scientists would have to regularly record, over a few years, the exact arrival time of pulses from several pulsars widely distributed across the sky, to know exactly how many times the pulsar has rotated between observations. As they look for any tiny variations in the time of arrival of the pulses, they would also search for the tell-tale signs of gravitational waves: correlated (linked) delays between pulses from neutron stars thousands of light-years apart.

What had been a completely theoretical concept became a more realistic proposition in 1982, when Don Backer discovered the first millisecond pulsar. Unlike typical pulsars such as those that Jocelyn Bell found, which spin roughly once per second, millisecond pulsars spin hundreds of times a second. Backer realized that this feature made it possible to measure the frequency of millisecond pulsars much more precisely, down to a few tens of nanoseconds, and this precise timing would then make them a much better bet for one day spotting a gravitational wave.

So how does a timing array work?

We first briefly encountered pulsar timing in Deeper Dive in Chapter 2. Assume we have a set of fifty different pulsars, which astronomers, over a decade, have observed very precisely: the arrival of each pulse has been measured to at least one thousandth of a single rotation over the very large number of total rotations in that ten-year period.

For example, say a pulsar is rotating at five hundred times per second, it will complete about 160 billion revolutions in ten years. The sheer number of very exact measurements is what gives the technique such enormous precision and power—if suddenly pulses from some pulsars arrive early, then late, and then early again compared to their expected arrival time, then it's clear that something has jiggled the person taking the data. If it's not possible to explain the change with some nearby noise (like an earthquake or a rumbling truck), then it might be a ripple of a gravitational wave washing over the Earth. If this shifting to earlier and later were to affect all the pulsars in a specific pattern calculated by Hellings and Downs, it would be the "signature" of gravitational waves squashing and stretching the space between the Earth and the pulsars, along the path that the pulses follow.

In a way, it's very similar to LIGO and Virgo—with the Earth and the pulsars being the equivalent of the test masses at the end of each arm of the detector. But with pulsars, it's necessary to take the data over many years, because the wavelength of these gravitational ripples is so long.

Backer wanted to do just that. In the early 2000s, he got together with colleagues at UC Berkeley, as well as astronomers David Nice of Lafayette College in Easton, Pennsylvania, and Ingrid Stairs of the University of British Columbia, and applied the Hellings-Downs theory to begin timing a handful of millisecond pulsars very precisely, using both the Arecibo Telescope and the GBT. Their aim was to spot gravitational waves—or at least to put limits on them, establishing a window of measurements within which they thought a gravitational wave might one day be observed. "Quietly, five of them started timing pulsars. They were doing it as kind of a small group, it wasn't a huge effort," says Ransom.[21]

But others found out, and were intrigued. In Australia, Dick Manchester liked the idea so much that he applied for—and received, in 2003—a big grant from the government to start pushing the technique on a bigger scale. As a result, he "got plenty of time on Parkes and money to hire postdocs and basically started the Parkes Pulsar Timing Array," or PPTA, says Ransom. Manchester was hoping that within five years the array would detect gravitational waves. This optimistically short timescale motivated other astronomers around the world to take seriously the quest of using pulsar timing to find gravitational waves.

European astronomers joined in at around the same time, linking the Westerbork Synthesis Radio Telescope, the Effelsberg Radio Telescope, the Lovell Telescope, the Nançay Radio Telescope, and the Sardinia Radio Telescope into the European Pulsar Timing Array (PTA). In the United States, Ransom got swept up in the excitement, too. "We knew that we had the two best telescopes in the world—Arecibo and the GBT," he says. So he and a dozen other American pulsar astronomers got together for a meeting. "We brought in Don Backer and his little group, and we said, 'Hey, you guys are doing this little project. But here's the Australians, they're doing this full force, they're doing this huge job. Let's do this right, we're here to work with you,'" Ransom recalls. They agreed to collaborate—and a few years and grants later, in 2009, NANOGrav (the North American Nanohertz Observatory for Gravitational Waves) was born. This North American pulsar timing array now routinely times more than seventy pulsars.[22]

These three efforts have been going on for more mode than a decade, but so far none have detected gravitational waves. Still, timing these pulsars regularly yields surprises. One day in 2012, Ryan Lynch, then a postdoc at McGill, was sifting through the data of a pulsar survey that the GBT had been running for a decade. He stumbled upon a millisecond pulsar, now called MSP J0740+6620—and after a few months of monitoring it, realized that it was an incredibly precise clock. With a group of colleagues, among them Scott Ransom, Lynch started timing it using NANOGrav. Soon Lynch noticed that the pulsar might have a partner star—allowing postdoc Emmanuel Fonseca and graduate student Thankful Cromartie to get the mass of the companion and then of the pulsar. Cromartie led the efforts, and the pulsar turned out to be the most massive neutron star detected yet, at a whopping 2.14 solar masses.[23]

But no gravitational waves—yet. Astronomers still don't know whether there are any supermassive black holes that are coming really close together and are about to merge, even though they have detected galaxies with two active galactic nuclei—evidence that two galaxies have merged and that their black holes are starting to waltz toward each other. "It'll take millions of years for them to merge, maybe even hundreds of millions of years," says Ransom. "But throughout the whole Universe there are galaxies that are merging all the time."[24]

Researchers are not about to give up, though—over the past decade, they have learned a lot more about how galaxies collide. They have also made great progress in developing simulations that eliminate from the calculations more of the noise and effects of the interstellar medium, all steps that help scientists to better describe and understand the Solar System's motions. Ransom estimates that within this decade, astronomers will finally detect gravitational waves from supermassive black holes using pulsars.

The first goal of pulsar timing arrays is to detect background gravitational waves from the mergers of supermassive black holes that have taken place throughout the evolution of the Universe. It's a difficult task: Imagine you are in a crowded room and everyone is talking. Attempting to detect a specific gravitational wave is like trying to hear a particular conversation across the room. With all the background noise all you can make out is a murmur. That murmur is exactly what pulsar timing arrays are trying to spot—the combined signals of thousands or even millions of supermassive black holes that are in the process of merging. "This is something that takes a really long time—we're not going to see what LIGO sees, where you actually see the merger," says Ransom. "We're seeing these black hole systems, thousands or even tens of thousands of years before they merge. But when they're that close, they're giving off these gravitational waves, these periodic ripples in spacetime. And when you add all of those together throughout the whole Universe, that's what this background is that we're trying to see."[25]

Once astronomers do detect such ultra-long gravitational waves, it should be possible to tell how many supermassive black holes are out there, how often they merge, and why. Is what's bringing them together just gravity? Or is there perhaps a lot of extra gas or stars near the black hole that cause them to merge faster than we would expect if only gravity were involved?

Meanwhile, other instruments are about to join the pulsar-powered hunt for ultra-long gravitational waves. In 2034, the European Space Agency plans to launch the first space-based gravitational waves detector, LISA (the Laser Interferometer Space Antenna). While LIGO detects rapid gravitational waves from stellar-mass black holes whipping around during the last seconds before they collide, and pulsar timing arrays hope to see supermassive black holes rumbling around in orbits that take years, LISA will explore the range in between: orbits that take minutes. "We expect it to detect merging white dwarfs and the actual mergers of the types of supermassive black holes that NANOGrav will detect in the early in-spiral stage," says Lynch. Astronomers know that there are tens of thousands of white-dwarf systems orbiting each other this quickly, but LISA should also see other interesting phenomena like small black holes spiraling into much larger ones. To detect slow, long-wavelength gravitational waves from all these events, LISA needs very long arms and an environment free of earthquakes and pizza trucks, so it will use of a set of three satellites trailing the Earth around the Sun.[26]

On the pulsar timing front, MeerKAT has already kicked off MeerTIME, the project mentioned earlier that will time a thousand pulsars. Scientists believe that the Thousand Pulsar Timing Array should give the hunt for ultra-low frequency gravitational wave detection a major boost, because MeerKAT is the most sensitive telescope in the Southern Hemisphere used for timing. (Arecibo is more sensitive than MeerKAT, and the GBT is close, while the new Chinese FAST telescope is more sensitive than all of them—but they are all in the Northern Hemisphere.) MeerKAT may be especially useful for timing pulsars that don't pulse regularly, but occasionally glitch or show variations in their radio and x-ray emissions. Right now there are only a handful of pulsars that are known to be extremely well-timed—mostly thanks to how long astronomers have been taking data. MeerKAT will be able to greatly improve the number of known well-timed pulsars and do so in a much shorter period, because its sensitivity will mean it's possible to get the same result much faster.[27]

Even if astronomers don't end up detecting gravitational waves with pulsar timing arrays, these arrays could be useful for other applications—such as making a pulsar timing-based timescale. The best time standard on Earth is International Atomic Time, based on atomic clocks. But are these clocks always right? "We've got a bunch of atomic clocks around the Earth that all average together to get time. But how do you compare it?" asks George Hobbs, an astrophysicist at CSIRO in Sydney. "If you've got the best clock, you can't compare it to anything." In 2012, he and his colleagues put forward a proposal to use the first pulsar-based time scale—the Ensemble Pulsar Scale (analogous to the free atomic timescale, Échelle Atomique Libre)—which is based on observations from the PPTA, to compare atomic time with pulsar signals. In 1996, atomic time was found to be slightly different from pulsar time, by just a few microseconds; showing, says Hobbs, that it's possible to use pulsars as an independent cross-check on International Atomic Time.[28]

Besides potentially being galaxy-sized gravitational wave detectors, as we know from LIGO, pulsars are also great at generating gravitational waves themselves. And that in turn makes them amazing tools to test whether Albert Einstein's most famous theory of gravity—general relativity—is correct.

Was Einstein Right? Testing, Testing . . .

Saying adieu is never easy—especially when it's to the work you have spent years on. But that's exactly what happened in September 2017, when Miguel Zumalacárregui, a theoretical physicist at the Berkeley Center for Cosmological Physics, came to the Institute for Theoretical Physics in Saclay, near Paris, to give a talk. It was still a few weeks before the LIGO and Virgo collaboration officially announced to the world the first-ever observation of a merger between two neutron stars. But with a collaboration of more than a thousand researchers, it's tricky to keep news under wraps; already, rumors about the collision were spreading across social media, and from university to university and conference to conference. And here was Zumalacárregui, ready to tell his audience that this detection—if true—would kill off a whole bunch of so-called alternative theories of gravity.[29]

Albert Einstein's general theory of relativity has been widely accepted for more than a century, replacing Isaac Newton's theory of gravity. In general relativity, gravity is not a force but a curvature of spacetime, so it predicts that massive bodies warp the fabric of spacetime, making light curve or bend when it passes near it. But scientists know that there's something wrong with general relativity: there's no way to reconcile it with the realm of the very small, that is, quantum mechanics. General relativity seamlessly describes gravity, such as why a pen falls down when dropped from a desk. It explains why if you're in an elevator, there is no way to know whether you are on Earth or in space, being pulled forward at a constant acceleration. This is called the equivalence principle—which states that gravitational fields are indistinguishable from accelerating frames of reference. The same way, general relativity explains how planets orbit a star and how galaxies collide. But zoom in, and we need quantum mechanics to explain how electrons orbit an atom, why an atom of plutonium decays, or why two atoms of hydrogen fuse to form helium, converting some of the hydrogen mass to energy—the process that powers stars before they exhaust all their fuel and die.

Quantum mechanics and general relativity don't work together: whenever we try to explain the small with general relativity, or the big with quantum mechanics, everything breaks down. We know general relativity isn't the right theory, so researchers keep testing it in various systems and environments, trying to find flaws. They also hope that by exploring alternatives, they can get closer to what that correct theory is. Alternative theories of gravity typically try to deal with two fundamental issues of general relativity that stand in the way of our understanding of the Universe. The first is the need for dark matter, the invisible matter that Fritz Zwicky believed was holding galaxies in the Coma cluster together. The other issue is that we need dark energy to explain the behavior of our Universe. The present understanding of dark energy is that it is some undetectable substance whose pressure makes the Universe expand. Alternative theories try to make it arise as a force that is part of how gravity works.[30]

Maybe, argue some theorists, gravity is more complicated than Einstein said, and we haven't found the situations where those complications become large enough to matter (unless dark matter and dark energy are those situations). That's why they are developing alternative theories to modify general relativity, in the hope that they will be able to describe all our observations seamlessly and without the need for any of the dark stuff. "There are also good reasons from the pure curiosity side: roughly a third of the unknown problems in physics—including dark matter and dark energy—have to do with gravity, such as fully understanding black holes, reconciling the quantum realm with gravity into a theory of quantum gravity, and the cosmological constant problem," says Zumalacárregui.[31]

Even before the LIGO detection, astrophysicists knew that a merger of two neutron stars would deal a blow to some alternative theories of gravity. But, says Zumalacárregui, at the time of his talk a lot of researchers still hadn't realized that their own pet theory was about to be thrown off a cliff for good. And so at Saclay he patiently explained how the rumored LIGO discovery would rule out many general relativity alternatives. After he finished speaking, plenty of hands shot up—with fellow scientists asking him in disbelief whether this or that specific theory was really, really dead. The conference was like a funeral, he says, where to some attendees he was the first to break the news about the person in the coffin. Still, Zumalacárregui adds, some of the reasons for modified gravity theories arise simply due to insufficient understanding. "Remember that before Einstein, many of the brightest scientists were working hard to reconcile electromagnetism with the ether, a completely wrong approach," he says. Indeed—ether was once thought to be a medium necessary for the propagation of light, but later it was confirmed that it didn't exist.[32]

Not surprisingly, perhaps, some scientists are wondering whether Einstein and general relativity are simply wrong. But it's not just about the dark stuff. For modified gravity theories to work, they need to reproduce all the standard predictions that general relativity so successfully makes. As it happens, most don't. Because many of their predictions turn out to be very wrong, researchers have to come up with error corrections and screening mechanisms that hide the wrong prediction on scales where it's possible to test them.

Perhaps the best-known alternative gravity theories are modified Newtonian dynamics, or simply MOND. These theories eliminate the need for dark matter by changing the definition of gravity—they assume there are two kinds of gravity and not just one. Sometimes, they say, the force of gravity is strong, making objects obey Newton's law of gravity—which states that the gravitational force between two objects shrinks according to the square of the distance that separates them. But, according to MOND, when the force of gravity is very, very weak—for instance, at the outskirts of a galaxy—this type of gravity decreases more slowly with distance, meaning it doesn't weaken as much. This would explain why the gravitational force of ordinary matter does not appear to be strong enough to keep rapidly moving stars inside their galaxies; and it does away with the need to inject dark matter into the model to increase the gravitational pull required to make up the difference.

MOND is just a broad idea. One attempt to make it into a fully fledged theory is TeVeS (tensor-vector-scalar), which adds a mathematical machinery that covers everything from the Solar System to black holes to galaxies to cosmology. As a form of MOND, TeVeS explains the velocity of stars in galaxies being the same everywhere—by making gravity stronger than Einstein or Newton would predict far away from the center. Then there are Galileon theories, part of a class of theories called Horndeski and beyond-Horndeski. Horndeski theories are named after Gregory Horndeski, the physicist who developed them in 1974. The wider physics community only really started paying attention to his ideas around 2010, after he had quit science to become a painter in New Mexico. Most of these theories try to eliminate the need for dark energy, trying to tweak explanations for the expansion of the Universe and the force of gravity. Other theories, like Brans-Dicke gravity and its variants, come from exploring philosophical or mathematical expansions of general relativity. There are also stand-alone theories, like that of physicist Erik Verlinde, who suggests that the laws of gravity arise naturally from the laws of thermodynamics, just like waves emerge from the molecules of water in the ocean.

Researchers who work on these theories usually have spent years developing them, working on the equations, and refining them. All of this work came to a screeching halt when the first neutron star merger was recorded. As LIGO's and VIRGO's mirrors shook ever so slightly, and the scientific collaboration went into overdrive to observe the first such collision ever, the space-based Fermi observatory also saw a short gamma-ray burst coming from exactly the same spot—which arrived on Earth just 1.7 seconds later. This proved that gravitational waves travel at the speed of light—pitilessly murdering TeVeS theories, which only work if those speeds are different. LIGO also killed off several Galileon theories, which had required an extra field to explain the Universe's accelerated expansion and required gravitational waves to move slower than light.

Not all theories got killed off: some Horndeski and beyond-Horndeski theories have survived, because they don't require a change in the speed of gravitational waves; also clinging on are some so-called massive gravity theories. Typically, physicists assume that the particle associated with gravity—the graviton—has no mass. But in these theories, it does, albeit a very small one, so it wouldn't necessarily move at the speed of light.

Still, while the collision of two neutron stars signaled a very prompt send-off to some alternative gravity theories, researchers have also been eliminating them in other ways—more meticulously but just as mercilessly—first using the bodies in our Solar System, and more recently, using pulsars. So far, they have been proving Einstein right, over and over again.[33]

But the more enormous the density and the stronger the gravity—with pulsars, it's as strong as it can possibly get without the star collapsing further into a black hole—the more chance there is that general relativity may break down. So scientists are tirelessly looking for the tiniest discrepancies in pulses to see if whatever pulsar system they are subjecting to a test is still consistent with Einstein's predictions. The results can also be compared to alternative gravity theories. Because the requirements are stricter, pulsar tests of general relativity allow physicists to draw more conclusions—and rule out some alternative theories.

Typically, astronomers use a pulsar in a binary system, say with a white dwarf or another neutron star, as a test mass with a very accurate clock attached. Then they calculate pulsar orbits very precisely by carefully timing the arrival times of pulses. Imagine you measured a pulse arrival time now, and one in ten years, to a microsecond: because we know exactly how many turns the pulsar made between those two measurements, we can calculate the pulsar's spin frequency to an accuracy of one microsecond in ten years—one part in 3×10^{14}, or 300 million million.[34]

Astronomers then can analyze how the pulse arrival times vary as a result of the radiation (such as, in this case, radio waves) traveling past the pulsar's partner star. They do this by measuring the Shapiro delay, that is, the time delay due to gravity's ability to bend light. "We are really good at measuring time," says Anne Archibald, an astronomer at the University of Amsterdam, and tells a story. When the Cassini space probe passed behind the Sun on its way to Saturn, scientists returned radar signals from it and measured their travel times very precisely. Sure enough, because the light was bent into a curved path, it took a little longer to reach them than it would have if the Sun's gravity wasn't in the way. This was the Shapiro delay, and the Cassini measurement is still the best measurement of it. The delay serves to constrain a different aspect of alternative theories of gravity—how they say light should be bent. And, of course, it allows astronomers to calculate the masses of the companion star and the pulsar itself.[35] (For more on the Shapiro delay, see "Deeper Dive: Kepler's Laws and Beyond.")

There are other general relativity effects, too. In 1998, Michael Kramer, who was then a postdoc and is now a director of the Max Planck Institute for Radio Astronomy and a guru on gravity tests, was observing the Hulse-Taylor pulsar with the hundred-meter telescope in Effelsberg. He noticed that the pulse shape seemed to be different from what had been published twenty-four years earlier. He searched for other examples of its pulse profile in past publications and noticed that the pulse shape was changing much more than had been recognized before. "I dug deeper, analyzed more data and at the end of it, I realized that I had seen the first real evidence for a GR [general relativity] effect called geodetic precession, which was predicted for the system just after its discovery in 1974," Kramer says. Geodetic precession is when the axis of a gyroscope shifts slightly as it goes around a massive object because space is bent, making more than 360 degrees in that circle. As a pulsar rotates around its partner, the mass of the partner makes it sink into a well in the fabric of spacetime—in other words, the lighthouse beam is shifting slightly with time, and after a while will completely miss the Earth, making it impossible for us to detect the pulsar.

Kramer's precision was limited, since the hundred-meter telescope he was using is much less sensitive than Arecibo, which had been used to discover the pulsar. Arecibo was also the dish that Joe Taylor, one of the astronomers who found the binary, and Joel Weisberg later used to test general relativity. Still, Kramer made a prediction for how the pulse shape would continue to change over time—and was even bold enough to predict that the pulsar would disappear from our view between 2020 and 2025, based on his data. "Perhaps not surprisingly, not many colleagues believed my result, I think," he says. A few weeks later though, Kramer got an unexpected email from Taylor, who sent him a plot. After reading Kramer's paper, Taylor and Weisberg went to Arecibo and did a new measurement of the pulsar—and it turned out that their measurement of the pulse width was spot-on with Kramer's prediction.

"That was a rather sweet moment," Kramer says, smiling—and was enough motivation for him to continue working with pulsars and keep testing general relativity. It's physicists like him who are searching for ever more extreme environments to test gravity—in the hope that one day they may find a scenario where the theory's predictions break down and they can gain a new and better understanding of physics.[36]

In 2000, they got lucky when astronomer Marta Burgay, now at the Cagliari Observatory, spotted a unique system—the Double Pulsar (PSR J0737–3039A/B)—in the Parkes survey data from Australia. As mentioned earlier, it's the only known binary system where both objects are—or rather were—pulsating neutron stars. The system is so relativistic that it allows astronomers to test aspects of Einstein's theory of gravity that have never been subjected to scrutiny before.

That day in May 2000 started just like any other day in Burgay's career as a researcher. She was sifting through pulsar survey data obtained by the Parkes Telescope, taking a closer look at the Galactic Center, when she spotted a rapidly rotating pulsar in a binary system. At first glance, it seemed to be just another binary millisecond pulsar with a not-so-interesting companion, just like many dozens of similar binary systems. "The discovery was not a particularly wow event," recalls Andrew Lyne, who worked with Burgay at Parkes. As with any other binary, they started timing it—observing it at specific intervals to see how it behaves. The researchers soon realized that the companion was likely another neutron star, just like the Hulse-Taylor pulsar. Yet it was a very tight binary system that made Einstein's general relativity effects very pronounced: these two massive bodies were curving spacetime in a big way. Burgay and her colleagues submitted the paper about the new binary to *Nature*.[37]

Four years passed. One day, astronomer Duncan Lorimer of West Virginia University wanted to test some new software and thought the newly discovered binary would be ideal for that. It was spring 2004 and he was using Parkes to run his test by gathering fresh data from the binary. Suddenly, Lorimer detected very specific periodic pulses, meaning that both stars were producing beams of radio waves. Straight away, he called Lyne, who was on vacation at the other side of the world, in Wales. Lyne remembers the moment his phone rang: "It was Duncan, calling to say that he found a three-second periodicity—and asking what to do about it."[38]

One of the two pulsars was rotating at millisecond speeds, completing forty-five turns each second, and the other one was spinning relatively slowly, once every 2.7 seconds. Why was the three-second periodicity so important? In a double star system, when the second neutron star forms, it behaves like an ordinary pulsar—it cannot accrete anything from its very dense companion. It spins down and turns off in a relatively short time, leaving only the long-lived, millisecond, "recycled" first pulsar. The three-second period was a hallmark of an "un-recycled" pulsar. "We saw the periodicity was indeed varying in just the way you would expect it if it was in orbit with this millisecond pulsar," Lyne says.[39]

This binary is subject to incredibly strong gravitational forces as both neutron stars whip around each other in tight orbits of just 2.4 hours, which in cosmic terms is incredibly close to each other. The system is also very nearly edge-on—we are almost in the plane of the orbit, which means that as seen from Earth, one pulsar passes almost exactly in front of the other (makes a transit). In March 2008, the slower pulsar disappeared; its pulse is still undetectable. The disappearance is another demonstration of geodetic precession, and researchers predict that its pulses should become visible again in 2035.

Using the system, scientists were able to test general relativity like never before. Before the beam from the slow pulsar vanished, astronomers could observe how the two neutron stars were orbiting each other, with a period of once every 2.4 hours. Occasionally, as the slower pulsar passed in front of its companion, its strong magnetic field would block the beam of radiation of the millisecond pulsar for thirty seconds or so. Because a pulsar is very small, only some 20 km (12.4 miles) across, it's not its physical mass that blocks our line of sight, but the cloud of intense plasma that spins with the slow pulsar and generates its own radio pulses. This cloud of plasma, held by the slow pulsar's magnetic field in a doughnut-like shape, rotates with the pulsar. Astronomer Rene Breton of the University of Manchester was able to measure and model the way the doughnut erratically blocked our view of the fast pulsar. As it turned out, the axis of rotation of the doughnut was precessing. As predicted by general relativity, the beam from the millisecond pulsar was arriving on Earth microseconds later, because its radiation had to pass near the second pulsar instead of traveling straight to us—which resulted in the previously established Shapiro delay and the now-measured precession. The test agreed with Einstein's predictions to within 99.99 percent.[40]

"I remember at one of our group meetings, Rene had a statistical question—there were two ways to analyze the data, he said, and one of them gave a result that agreed with Einstein and the other disagreed, and he wasn't sure which method was right," says Archibald. "'Wait,' our supervisor Vicky Kaspi said, 'tell us about the methods but don't tell us which is which.' So we argued about it and settled on one method as the right way. And sure enough, it was the one that agreed with Einstein—but she taught us an important lesson about the importance of controlling our biases."[41]

Kramer and his team now have data that suggest that the orbit of the two pulsars shrinks every day by seven millimeters—the indirect evidence of gravitational waves. The Double Pulsar is much smaller than the Hulse-Taylor system (which Joel Weisberg continues to monitor to this day)—and the orbital shrinkage continues to follow the pattern predicted by Einstein's theory, that is, it is losing energy by turning it into gravitational waves much more quickly.[42]

AND THEN THERE IS THE UNIQUE TRIPLE SYSTEM.

In the summer of 2006, engineers at the Green Bank Telescope in West Virginia noticed that something was wrong with the track they were using to move the machinery left and right, in azimuth. It was wearing out, because GBT's weight on completion—about 7.7 million kg (17 million pounds)—was more than originally planned. GBT's administrators decided to tear out the old track and install a new one, a huge undertaking that would take the entire summer. Because during this time the telescope couldn't move in azimuth, it couldn't track sources across the sky; it was effectively shut down for normal operations. But, as with Alex Wolszczan's pulsar survey while Arecibo was being repaired, pulsar astronomers—among them Archibald and Lynch (who later became an associate scientist at Green Bank Observatory)—realized that it would be perfect for a sky survey. They reckoned that if they used the low-frequency receiver, each night they could pick an elevation and just let the sky drift by; this gave them about two minutes on any piece of sky. Since nobody else needed the telescope, they were able to use it for almost continuous observations for a couple of months, gathering 120 terabytes of data—a survey called the "GBT 350 MHz Drift-Scan Survey."

The survey ended when the telescope was fixed, and they set to sifting through all their data for pulsars. That took them several years. They had already found a few dozen new pulsars when in 2013, as they got close to completing their work, they found something unusual on the very last disk holding their data.

They discovered a binary, some 4,200 light-years away—a pulsar and a white dwarf. So far, nothing particularly unusual. But as they tried to calculate the motions of this binary system, they realized that they could not predict the orbit of the pulsar if they assumed that it was just going around another star.[43]

There was simply no solution to describe the orbit that way. "It dawned on us that there's something weird going on," says Lynch. Then Ransom had an idea—he saw that it looked like the orbit of the pulsar was systematically shifting and suggested that there could be another star influencing the orbit. That it was a triple system. He sent an email to Lynch early in the day, and later to the whole survey collaboration.[44]

Ransom then led the efforts to verify his hunch, together with Ingrid Stairs, Hessels, Archibald, and Lynch.[45] And indeed, when they introduced a third body—another white dwarf—into the model, orbiting the pair, everything fell into place. Such a system, consisting of a millisecond pulsar with two stellar-mass companions, had never been observed before. "At that point, we didn't really have any way to deal with a triple system," Archibald says.[46]

Archibald was finishing her PhD thesis at the time—on another source they had found in the same survey—and needed a break from spelling corrections. She went through her email and saw colleagues

Anne Archibald
(Courtesy of Anne Archibald)

complaining that they didn't know how to properly time a triple system. "I thought, 'but couldn't you just . . .' and, at about four in the morning, I started coding the direct integration approach we still use," she recalls. It wasn't actually the first time she'd written code to directly integrate an n-body system, a system with more than two objects. At the age of fifteen, she read the science-fiction book *Ringworld*. "In it, there is an alien race that lives on a set of five planets arranged to orbit their common center of mass. 'Is that stable?' I wondered. So I wrote some code to simulate the motion. Of course, at fifteen I didn't know what I was doing and the integration code was itself unstable, so I didn't really get an answer. But it was clear that implementing an n-body integrator need not be difficult."[47]

Lynch says that with the Triple System, "Anne Archibald did fantastic work really coming up with numerical solutions that allow you to perfectly describe the orbit of all three of those stars." Her efforts paved the way to using the system to test general relativity in a very special way. The researchers realized that the three bodies exist in a very tight space, smaller than Earth's orbit around the Sun—which is great for testing an aspect of general relativity called the strong equivalence principle.[48]

The equivalence principle is the cornerstone of Einstein's general theory of relativity. It states that gravity affects objects equally, regardless of their composition or mass; in other words, any two objects regardless of their masses or makeup will be affected by a gravitational field in exactly the same way. For most purposes, physicists work only with a less strict version called the weak equivalence principle: it says that objects fall exactly the same way, regardless of composition or mass, but only if their gravity isn't too strong. A famous example is that if one removes the resistance of air and drops a feather and a hammer from the same height, they will hit the ground simultaneously. This experiment has been successfully demonstrated not only on Earth, in NASA's huge vacuum chamber, but also on the Moon.

The strong equivalence principle takes the concept of the weak equivalence principle but adds one condition: even objects with their own gravity should fall the same way. An object like a planet or star is held together by its own gravity, and the gravitational energy that holds it together, according to Einstein's famous $E = mc^2$, has mass. Does this mass fall the same way as ordinary matter we encounter in our everyday life, say, a chair?[49] Does gravity fall the same as matter? "Weird as it sounds, this is actually a sensible question," says Archibald. "While Einstein's theory of general relativity says that gravity falls exactly the same way as everything else, almost all alternative theories predict that actually gravity should fall differently from matter. Put another way, if you have a thing with a lot of gravity holding it together, that strong self-gravity will change how the object falls."[50]

It's not possible to create an object with strong gravity in a laboratory. Until recently, the best test of the strong equivalence principle was in the Earth-Moon-Sun system. Gravity holds the Earth and the Moon are together as a pair, and the Earth's gravity is much stronger than the Moon's. So if Einstein was wrong and the self-gravity does affect how things fall, then maybe the Earth would fall differently from the Moon.

It's not possible to drop the Earth and the Moon from a tower, as the Dutch scientist Simon Stevin did with cannonballs from the Nieuwe Kerk (New Church) in Delft, but as the Earth and Moon orbit the Sun, they are continually "falling" toward it. This means that if they experienced different accelerations because of their different amounts of gravity, the Sun would be pulling the Earth-Moon system apart. "Only a little [data are available], but we have very precise measurements thanks to the reflectors that the Apollo astronauts and the Russian robot lunar landers left on the Moon. And sure enough, there doesn't seem to be a difference in how the Earth and Moon accelerate towards the Sun," says Archibald.[51]

WHILE IT'S NOT NECESSARY to have three bodies for the test, there has to be some external gravitational field. Astronomers have done gravity tests with pulsars, using, for example, a pulsar and a white dwarf experiencing the pull of the gravitational field of the galaxy—"falling" into the galaxy, so to speak. But the problem with such tests is that the galaxy produces a really tiny acceleration on the pulsar system—it's not pulling very hard, so a difference in how it affects the two bodies is also tiny.

The good thing about pulsars, though, is that they are extremely dense objects—just like white dwarfs and black holes—and their binding energy is much stronger. For a pulsar, for example, it's 10 to 15 percent of its mass. So if that gravitational binding energy experienced gravity differently from matter, the effect would be much more visible in a neutron star system.

The problem is that while we know quite a lot about our Solar System, we don't necessarily know the mass of a neutron star or its companion. "We have to first know how Einstein said it should fall, which is tricky if we don't know exactly how strong the external gravitational pull is," says Archibald.[52] For example, when astronaut Dave Scott stood on the Moon and dropped a hammer and a feather in 1971, he let them both fall in the same gravitational pull—and they fell the same way. With such tests, it is necessary to have two objects falling in the same gravitational pull external to them both. In the Earth-Moon-Sun system, the gravitational pull is from a third body nearby, that is, the Earth and the Moon are being pulled (they are "falling") toward the Sun.

This need for an external gravitational pull meant that the Triple System "was really kind of a unique system for probing this equivalence principle in ways that we can't do in the Earth-Sun system, or the Moon-Sun system, or with pulsars that are just kind of out there in the gravitational field of the galaxy," says Lynch.[53]

Archibald and her colleagues did the test using their computer model with the exact parameters from their observations—and sure enough, the pulsar and the inner white dwarf do fall at exactly the same rate in the gravity field of the outer white dwarf, despite the neutron star being much, much more massive than its binary companion. They built a model that allowed for the strong equivalence principle violation, with a certain parameter called Delta. According to Einstein's theory, Delta should be exactly zero. "We didn't just take the parameters from a normal orbit and try to look for deviations. We actually fit our model with its maybe-not-Einstein physics to our measurements directly," says Archibald. The researchers simulated a lot of orbits with strong equivalence principle violations, but the ones without such violations—those that obeyed Einstein—fit their data better. Their best-fit value of Delta wasn't exactly zero—there was some noise—but Delta was zero within their margin of error. It turned out that the Triple System behaves just the way Einstein predicted, with deviations as small as three parts in a million showing up in the test.[54]

Alas, their calculations also eliminated those alternative-gravity theories that imply that the two wouldn't fall in a similar manner. Einstein was proved right—yet again.

What's next for general relativity and alternative theories of gravity? This question, Zumalacárregui says, keeps him up at night more than he'd like. But at least now, with many theories killed off, it's possible to focus on the few survivors and make them better. In the future, the Square Kilometer

Array and other next-generation telescopes will hopefully find more unusual pulsar systems, which would allow researchers to do even better tests of general relativity using pulsar timing. LISA should help tremendously too, once it launches.[55]

Still, Archibald says, theorists are likely to keep coming up with new theories no matter what. One discovery that astronomers are hoping to make in the near future may probe general relativity even better—spotting the first pulsar–black hole system, perhaps by using an existing telescope like the GBT, Arecibo, or MeerKAT, or perhaps by some new telescope that is just coming online or planned for the future, such as FAST or the SKA. Gravity is strongest in black holes, so a pulsar orbiting a black hole could be the ideal test mass to analyze the black hole's spacetime. This would help us to learn more about black holes, too, for example by measuring their spin. A project called BlackHoleCam is working with the Event Horizon Telescope to combine the data from supermassive black hole imaging and pulsar timing. They hope to one day spot a pulsar orbiting a black hole, ideally around Sgr A*. And if we spot more than one, it could also settle the question of whether dark matter or pulsars are the source of the gamma-ray excess in the Galactic Center (see Chapter 6). Very soon, neutron stars may indeed help us to either knock Einstein's general theory of relativity from its pedestal or say adieu to all alternative gravity theories for good.[56]

DEEPER DIVE: Kepler's Laws and Beyond

When astronomers monitor a pulsar and find that its orbit follows Kepler's laws of planetary motion, which describe how planets orbit the Sun, they usually define it with so-called Keplerian parameters. But if there are deviations, researchers make the Keplerian orbits a bit more flexible by adding what's known as post-Keplerian parameters. These tweaks help them describe what researchers see.

One such parameter is the Shapiro delay. Pulsars' companions such as white dwarf stars are massive bodies, so they also bend light—or in this case, radio waves—emitted by the pulsar. According to Einstein's theory of general relativity, the radiation should go down the well as it passes the companion and then up again—thus arriving at our telescopes later than if it were traveling in a straight line, as it would do if spacetime were flat. Shapiro delay does not test general relativity but it is predicted by it—and if one assumes that Einstein is right, general relativity is used to convert the signature of Shapiro delay into a measurement of the mass of the companion. Once this mass of the companion is known, it's relatively easy to get the mass of the pulsar.

Unlike the Shapiro delay, most gravity tests are based on measuring changes in an object's orbit—usually tiny variations that add up over time. For instance, as the pulsar and its companion travel around their common center of mass, their common orbit should shrink, bringing them gradually closer together on the fateful dance to collision as the system loses orbital energy due to the emission of gravitational waves. That's what Hulse and Taylor first showed in 1974, and LIGO's direct detection of gravitational waves from the GW170817 merger was crucial direct evidence—and an important test for general relativity.

Another relativistic effect (and post-Keplerian parameter) that astronomers try to document in binary systems is the advance of periastron—the fact that the orbital elliptical shape, which would be a perfect ellipse under Newton, doesn't close back on itself according to general relativity. It shifts a little bit each orbit. It's the same phenomenon as the advance of the perihelion of Mercury—as Mercury orbits the Sun, it doesn't follow exactly the same path each time; instead, its perihelion, the point on its orbit when it's closest to the Sun, advances forty-three seconds of arc per century because of the curvature of space.

Measuring the advance of periastron in neutron star binaries—in other words, how quickly the orbit is shifting—allows astronomers to calculate the pulsar and companion masses, again assuming that Einstein was right. But if Einstein was wrong, these formulas are wrong, and researchers might get different masses out of the two methods. Shapiro delay is an effect of general relativity outside of Newton's gravity laws—and Einstein predicted that all the deviations that don't work with Newton's theory should be related in a specific way, which in this case means that they should all agree on the same masses for the pulsar and the companion. "If they don't, something is wrong with Einstein's theory," says Archibald.[57]

Once two post-Keplerian parameters are known, if you then measure a third one, it becomes possible to check whether it agrees with the value predicted from the other two, once again using Einstein's theory. Add a fourth parameter, and you get yet another check for Einstein's theory. If scientists were to find any measurable deviations from these predictions, then Einstein's general relativity may not be the best theory of gravity after all—but so far, no measurements of post-Keplerian parameters have managed to knock down Einstein's equations.

From a general-relativity point of view, most pulsar systems are somewhat boring; they won't allow astronomers to observe any post-Keplerian parameters. Just a few neutron star systems are "relativistic," meaning that they are massive enough and move around their orbits quickly enough for general relativity to produce observable effects. The Hulse-Taylor binary is one. Others include the binary that Alex Wolszczan discovered in 1990 together with the pulsar orbited by three planets that he and Dale Frail first documented in 1992.[58]

9

Fast Radio Bursts, an Unfinished Chapter

"DECLINED."

There were other words in the email, but this one stood out—as if written in bright, bold red. It was June 2007 and the first time that Duncan Lorimer, an astrophysicist at West Virginia University, had submitted a paper to the prestigious academic journal *Nature* as the lead author. Like most scientists, Lorimer had always dreamed of publishing a first-author *Nature* paper, and he was certain this was his chance. But his manuscript wasn't even sent to referees—the British journal's editorial staff judged the paper "unlikely to succeed in the competition for limited space" and not "of great topical interest to those working in the same or related areas of science."[1]

Lorimer felt the floor go wonky under his feet. Because of the rejection, one of the biggest astrophysical discoveries of the early twenty-first century—the existence of fast radio bursts (FRBs), the enigmatic, extremely powerful, and ultra-short flashes in space lasting just milliseconds that could help astronomers measure the composition and dynamics of the intergalactic medium, the strength of intergalactic magnetic fields, and shed light on other mysteries—nearly got overlooked.[2]

Fast radio bursts were once considered a rarity. Discoveries were few and far between, and because they were so brief, it seemed impossible to pinpoint their location. Scientists still don't know what they are and where they come from, although neutron stars have been put in the frame: the leading theory suggests that these flares might be generated by magnetars. What we do know is that these bursts are a billion times more luminous than typical pulsars, they can discharge as much energy as five hundred million Suns, and they come from far beyond our galaxy.

Only recently have scientists begun to localize—pinpoint the source of—FRBs, even those that only flash once. They are also detecting more of them than ever before. There are two reasons for these scientific successes: better telescopes and . . . aliens. Well, not exactly aliens themselves, but the search for them. Among the many astronomers and astrophysicists trying to understand the nature of these mysterious signals is an eccentric Russian-Israeli billionaire who, in his persistent search for intelligent life beyond Earth, has ended up co-financing one of the most complex and far-reaching radio scans of our Universe ever attempted.

Some astronomers used to think that aliens might indeed be behind the flashes, although that's unlikely. Beyond magnetars, other theories put forward by researchers suggest that FRBs could be the result of cataclysms such as supernovas, collisions between black holes and neutron stars, or maybe dark matter in the form of tiny, primordial black holes giving off flashes when they collide with stars. Or a more mundane reason: that the signals aren't from space at all but from microwave ovens here on Earth.[3]

It all started when Lorimer's undergraduate physics student David Narkevic came to meet his adviser for their weekly one-on-one in early February 2007 at West Virginia University, just beyond the banks of the Monongahela River that cuts through the city of Morgantown. "David is a type of a person who doesn't have the highs and lows in terms of emotional state," says Lorimer, jokingly. "So he came in and it just seemed like any normal meeting. We just started talking about what he's been up to that week." A few minutes into the chat, Narkevic pulled out a plot, with an ultra-tall peak that rose way above any pulse usually produced by a pulsar. Lorimer's eyes opened wide. "It was like, wow, wait a minute, what's that?" recalls Lorimer. "It was really surprising; I just didn't know what to really make of it."[4]

For a few weeks before the meeting, Narkevic had been sifting through a mountain of archival data from the Parkes Multibeam Pulsar Survey of the Magellanic Clouds, the largest pulsar survey to date, which started in August 1997 and ran until 2001. Lorimer had asked Narkevic to search for a type of unusual and just-then discovered pulsar called a rotating radio transient (RRAT). Lorimer talks fondly about these "rats," partly because they were first found by his wife, astronomer Maura McLaughlin. She spotted them around 2004, one year after they got married. ("Our program for the wedding was written in the style of a scientific paper, describing how we met and stuff like that—but written in this sort of dorky, scientific way," chuckles Lorimer.)[5]

RRATs are different from most other pulsars because their emission of radio waves is very unpredictable. Even today they are an enigmatic population of pulsars, best detectable through single pulses rather than as steady periodic sources. Astronomers know of a number of pulsars that do both: emit periodically but also with sporadic single pulses. The most famous example is the Crab Pulsar, which once in a while emits a jet of radiation that is more than ten times the intensity of its periodic emission.[6]

By contrast, RRATs are usually spotted via their erratic single pulses, some that last merely a few milliseconds and are detectable for less than a second per day. "We know that it's arbitrary. But we quickly realized it as a new way to find neutron stars that complemented the previous periodicity searches," says Lorimer.[7]

Thanks to the study of giant single pulses, researchers have known for a while that it is possible to detect a powerful emission coming from outside our galaxy. Most pulsars, however, are fairly weak sources of radio waves, which means that with today's telescopes we usually are able to spot them only inside the Milky Way or in the nearby Magellanic Clouds.

Lorimer and McLaughlin wanted to push the boundaries a bit, hoping to find giant pulses from RRATs and other pulsars coming from farther afield. They thought that they should be able to detect them.[8]

And so here it was—a single giant pulse in Narkevic's data, which lasted just five milliseconds but was ten billion times more powerful than a pulse from a typical pulsar. The way it was dispersed, Lorimer straight away noticed that it seemed to originate from way, way outside our galaxy, somewhere in the direction of the Small Magellanic Cloud. Our galaxy is 100,000 light-years (or about 30 kiloparsecs) across, but when Lorimer and Narkevic looked at the dispersion measure, it was ten times greater, suggesting that the source was definitely not in our own backyard.[9] (For more on the dispersion measure, see "Deeper Dive: The Interstellar Medium, Home of Neutron Stars," in Chapter 2.)

Lorimer was certain the discovery was a whopper. He went home that day and excitedly told his wife that they might have stumbled upon a brand-new object in space. She was excited, too, but for the first few weeks, the two didn't really know what to make of it. "It was just so amazingly bright that I just kind of filed it away for a while," says Lorimer. They were overwhelmed—by the potential discovery but also at home; their first child, Callum, had just turned two, and Maura was expecting their second, Finlay, who arrived in July 2007. During this time, both Lorimer and Maura were continuing to teach classes and supervise undergraduate and graduate students. "I think it was like, Oh, yeah, that looks great, but let's put it away for now," says Lorimer.[10]

In April, though, he boarded a plane to Parkes. There, he met his former graduate adviser, Matthew Bailes, to look at the plot together. Not only was Bailes thrilled, but his excitement was so infectious that Lorimer was soon enthusiastically joining in.[11] Within days they started working on a paper. It was ready by June and included a plot of the signal—a now-famous peak that rose well above any known pulsars. The signal looked so stunning, says Bailes, that he had difficulty sleeping during that time. Time and again he thought about how amazing a discovery it would be if the signal was indeed that bright and had come from that far away.

But then came *Nature's* rejection. Lorimer was upset, as was Bailes—although Bailes had a somewhat thicker skin; after all, sixteen years earlier he and Andrew Lyne had had to retract the paper claiming that they had discovered the first planet orbiting a pulsar. "Forget *Nature,*" Bailes remembers telling Lorimer, to cheer him up, and they agreed to send the paper to *Science*. It was published in November 2007, no questions asked. Lorimer was relieved, but he needed to know: what object had produced this burst that would soon bear his name?[12]

ENTER AVI LOEB. Now the chair of the astronomy department at Harvard University, Loeb grew up in Israel, where for as long as he can remember, he has been fascinated with life—on Earth and elsewhere in the Universe. But he has also been annoyed that most scientists seem to shun the search for intelligent life in the cosmos and focus on trying to spot signs of primitive microbial life instead, searching for its chemical fingerprints in the atmospheres of exoplanets. In his view, SETI should be in the mainstream; instead, it's considered a fringe science—probably, he thinks, because of all the negative portrayals of aliens in science fiction and all the dubious reports from UFO spotters.[13]

It's not that other scientists have never seriously considered the possibility of intelligent life beyond Earth. After all, back in 1967 when Jocelyn Bell stumbled across the "scruff" in the data from the interplanetary scintillation array in rural Cambridgeshire, she also wondered whether aliens were saying hello. (As mentioned earlier, that's why, while the signal was still a mystery, researchers only half-jokingly named it LGM-1—for "little green men"—and the other three pulsars that Bell discovered were dubbed LGM-2, LGM-3, and LGM-4.)

What helped Bell resolve the alien conundrum was that she soon spotted a second pulsar, then a third and a fourth—from very different directions of the sky, which was enough proof that these were naturally occurring objects and not aliens.[14]

Lorimer wasn't so lucky.

Just after he and Bailes published their paper, in November 2007, there was interest all right, with researchers coming up with theories for the oddball burst. Not much later, Loeb happened to be visiting Melbourne. He had read the paper, was mildly intrigued, and wanted to chat with Bailes. Loeb thought the burst was enigmatic, but like many others assumed that it might have been an instrumental artifact.[15]

Then everything stalled. For the first six years after the discovery, neither Lorimer nor anyone else found any similar bursts. The "fluke" explanation persisted—and the fact that sixteen similar giant pulses at Parkes were later shown to be caused by the door of a microwave in the observatory's visitor center being suddenly opened during a heating cycle didn't help Lorimer's case. People were seriously doubting that the peak first spotted by Narkevic had been real. Even Maura McLaughlin wrote a paper concluding that her husband's discovery was probably based on a mistake.[16]

Around the same time, however, Loeb was casually exploring the possibility of intelligent civilizations beyond Earth and theorized how we might be able to spot them. One of his papers suggested that radio antennas made to detect the emission from hydrogen in the early Universe (atoms emit radiation at specific wavelengths, helping astronomers understand what type of gas existed in space in the past) could also detect radio signals from aliens as far as around ten light-years away. Another paper called for scanning the Solar System with the Hubble Space Telescope for artificial lights from aliens, while yet another explored how we might detect industrial pollution in the atmospheres of exoplanets. Lorimer, however, wasn't interested in alien worlds and didn't follow Loeb's papers. Instead he was hoping that scientists would spot at least one more burst that was as weird as the flash he had noticed.[17]

And at long last, it happened: in 2013, an international team published a paper led by Dan Thornton, then a PhD student at the University of Manchester. They had discovered four bright radio flashes in the year's worth of data from Parkes.[18] The scientists suggested that the flashes were likely taking place every ten seconds or so—or roughly ten thousand times a day. Following some deliberation, the bursts received a name: FRBs. (This after scientists discarded the idea of calling them fast radio transients, or FaRTs for short. Also, says Bailes, GRBs were "already a huge thing, and if we called them FRBs, we'd cash in on a similar acronym.") Fast radio bursts they were, more and more researchers the world over finally got convinced that they were real, and gradually, the enigmatic FRBs even found their way into physics classes at universities. Lorimer shot to fame—although at the time, there were more theories about the bursts' origin than there were FRBs themselves.[19]

Loeb knew of the developments but didn't want to get involved just yet. At a dinner in Boston in February 2014, however, he found himself talking to a charismatic businessman—about life, the Universe, and everything else. It was the Russian-Israeli entrepreneur and billionaire venture capitalist Yuri Milner. A physicist himself, Milner is well known in Silicon Valley through his high-profile investments in technology giants such as Facebook, Twitter, WhatsApp, and many others. Most of all, however, Milner has always been fascinated with life beyond Earth.

A year later, in May, Milner visited Loeb at Harvard. The businessman asked Loeb how long it might take to get to Alpha Centauri, the star system nearest to us. Loeb told him he needed half a year to come up with an answer and find the technology that might take people there in their lifetime. Intrigued, Milner asked Loeb to lead one of the projects he was about to announce: Breakthrough Starshot. It's part of a bigger, $100 million venture called Breakthrough Initiatives—the largest ever private-cash injection into the search for alien life, with five projects announced at a lavish reception in a London hotel in July 2015. I accompanied Loeb to the banquet, and he was positively glowing.[20]

One of the five projects is Breakthrough Listen, which has been championed by, among others, both the famous late astronomer Stephen Hawking and the British astronomer royal Martin Rees. The project uses many of the radio telescopes on Earth to spot signals that might come from extraterrestrial intelligence. Milner's money started to be rapidly invested into cutting-edge technology such as computer storage and new receivers at existing radio observatories including Parkes and the Green Bank Telescope. Astronomers were glad to accept the funds, whether they believed in aliens or not.[21]

And it was at that moment that for the first time a fast radio burst made an encore appearance.

IT WAS A SUNNY but chilly afternoon on November 5, 2015, and Cornell astrophysicist Shami Chatterjee had just sat down at his desk when an email popped up. He glanced at a rather boring subject line and at first didn't think too much of it: "A minor point of interest regarding the Spitler Burst." The email was from a graduate student at McGill University in Montreal named Paul Scholz, and was addressed to a mailing list of some forty people. All were astronomers analyzing data from the largest survey of the Galactic Plane to date called the Pulsar Arecibo L-band Feed Array (PALFA), undertaken with the Arecibo Telescope; Chatterjee was one of them.

Having first gone to get a coffee, Chatterjee clicked on the email a few minutes later . . . and his jaw dropped. "It's hard to overstate how shocking and surprising this email was," he says. It announced that one of the previously observed FRBs, officially called FRB 121102 and unofficially the Spitler Burst, had flashed a second time.[22]

This burst was one of just twelve FRBs that had been detected back then. It bears the name of Laura Spitler, an astronomer at the Max Planck Institute for Radio Astronomy in Bonn, Germany, who had spotted it back in 2014. She found it in old PALFA data; the Puerto Rico dish had actually caught the burst two years earlier, in November 2012. It was the only one at that time that had been spotted from somewhere other than Parkes, which to scientists came as a relief—it removed any lingering doubts that something might be wrong with Parkes.

Netting the repeating burst wasn't total serendipity. With so few bursts detected at that point, scientists had been hoping that they might repeat. After all, they might simply be a particularly powerful type of pulsar, and with multiple pulses from the same source, it's much easier to convince everyone that it's real. So in May and June 2015, the PALFA collaboration decided to monitor the sky in the direction from which Spitler's burst had come from, but using Arecibo, a much more sensitive dish than Parkes. For months afterward, using a supercomputer, Scholz had been combing the telescope data. And suddenly, there it was: a second flash, from the same direction.[23]

It was amazing, says Chatterjee. Back then, most physicists had assumed that these ultra-powerful FRBs simply would not be able to repeat; after all, none of the eleven flashes found before the Spitler Burst had. "Everyone *knew* that FRBs don't repeat, because it requires vast amounts of energy for a pulse to be visible across intergalactic distances, and that means an explosive origin of some sort—so no repeats," Chatterjee says. "And here it was—the only FRB we had found at Arecibo, and it was repeating." After he got over the shock, he ran down the corridor to tell a colleague, astronomer Jim Cordes, to look at his email, then came back to his office and started typing a response. Meanwhile, his inbox started to fill up. By midnight, the thread had fifty-six messages from various astronomers.[24]

Because Scholz was the sender, he felt that he had to take a guess on what the source of the repeater might be. Clearly, many previous models suggesting a cataclysmic origin for FRBs—at least for this one—were suddenly out, because a repeating burst can't be caused by, say, a supernova or a collision between two neutron stars. So he threw an idea out there: "extragalactic magnetar," a young, incredibly magnetized, rapidly spinning neutron star. Others in the thread chimed in, and Maura McLaughlin was the first person to agree with Scholz's guess.

But where was it? Yes, the source had flashed twice, but astronomers knew that to localize it, whatever it was would have to flash again—and ideally be seen by several telescopes at once so the signals could be compared. Immediately, Chatterjee and his colleagues submitted a proposal to use the Very Large Array in New Mexico for ten hours, hoping that its twenty-seven dishes would be just what they needed to determine the location of the burst. Science now turned into a waiting game. After scanning the sky for ten hours, checking every few milliseconds, they detected absolutely nothing. Chatterjee was bummed.[25]

In the meantime, Milner was getting impatient about his Alpha Centauri idea—and got in touch with Loeb asking for an update. It was the end of December 2015, recalls Loeb, and he was in Israel, getting ready for a weekend trip to a goat farm in the south of the country. Now, however, he was asked to immediately put together a presentation outlining his recommendations for technology to take humanity to Alpha Centauri. The next morning Loeb scrambled to find a place with an internet connection, next to the reception desk of the goat farm, and prepared a PowerPoint presentation explaining how humans could use a spacecraft powered by a giant light sail; two weeks later, he showed it to Milner at the billionaire's home in Moscow. With the light sail idea still at the top of his mind, Loeb heard of the Cornell team's detection of a repeating burst; it made him remember his nearly decade-old conversation with Bailes back in Melbourne about the first FRB. He was intrigued and thought he might have an idea about the origin of the flashes—something he would come back to several months later.[26]

Half a year passed. In April 2016, Chatterjee wrote another proposal asking for more time on the Very Large Array—forty hours—in the hope of spotting another burst from the repeater. PALFA folks got the time, but again, detected nothing. It was getting embarrassing, but stubbornly Chatterjee emailed the Very Large Array management yet again—begging for another forty hours in August 2016.

Everyone was losing hope, and the situation was becoming more and more like Lorimer's, who had been hoping for another FRB to prove that he wasn't seeing things. But Chatterjee didn't have to wait six years; at last, whatever was sending out these flashes decided to play nice. "It looks like the fast radio burst came out to play today," Casey Law, the scientist monitoring the Very Large Array in real time, emailed his colleagues. The repeater then flashed again and again, eight more times. Unlike with pulsars, the bursts were sporadic, flashing at irregular intervals, and once there was even a "double burst" of signals only twenty-three seconds apart. Chatterjee and his team managed to identify a source location: a dwarf irregular galaxy about a gigaparsec (just over three billion light-years) away from Earth. That's astonishingly far. And given the frequent repeating pulses, something out there was recharging its mega battery ultra-quickly, in less than a second. What could it be?

One model started gaining ground: Paul Scholz's original suggestion that it was a magnetar. Indeed, in terms of brightness, a magnetar outburst from that far away should be detectable on Earth. And the small home galaxy of the source is a likely place for a magnetar: it has a lot of star formation, and magnetars in general tend to form from a type of supernova called Type-I superluminous supernovas. These often happen in dwarf irregular galaxies, thought to be similar to the early galaxies that existed when the Universe was very young.[27]

Right after the Big Bang, the Universe was filled mostly with hydrogen, helium, and a bit of lithium. As early stars started to form, their novas and supernovas seeded the interstellar medium with the heavier and heavier elements produced through nuclear fusion during the stars' lifetime. Each new generation of stars had more and more of these "metals"—elements heavier than hydrogen, helium, and lithium—which increased the overall "metallicity" of the Universe. But astronomers think that dwarf irregular galaxies have low metallicity—that they have formed from hydrogen and helium that stayed behind from the early days of the Universe. This way, these small galaxies produce more massive stars with—it's believed—stronger magnetic fields, leading to magnetars when they die. And such a magnetar, researchers think, should be able to produce ultra-powerful outbursts like the observed FRBs.[28]

But generating such powerful bursts in such quick succession would make any magnetar quickly run out of fuel. That's why some researchers think that this could be a very young magnetar, probably less than a hundred years old. Such newborn magnetars have very intense magnetic fields that are unstable, and can undergo dramatic episodes of reconnection and reorganization, whereby the field moves from one quasi-stable configuration to another and releases a huge burst of energy. In our galaxy, we would see these outbursts as soft gamma repeater flares—in the x-ray or gamma-ray spectrum. "We don't see that here—but a possible model says that we don't see high energy emission because it is absorbed in the nebula," says Chatterjee.[29] The nebula is the expanding cloud of debris, gas, and dust that is left over after a supernova. As the magnetic field continually reconnects, it injects energy into the nebula, which releases sporadic but giant bursts. The shocks from these bursts, says Chatterjee, would produce radio flashes that can travel across cosmic distances. There is some supporting evidence for this idea: the repeating signals are streaming from the same direction as a steady source of radio emission, which could be the background signal from the nebula.

Still, there are problems with the magnetar model: there haven't yet been any FRBs from magnetars much closer to Earth, such as the one in our galaxy called SGR 1806 – 20 that produced a giant gamma-ray burst in December 2004, but no FRBs. Of course, it's possible that magnetars give off FRBs in narrow beams—and we would only be able to detect it if it's pointing directly at Earth.[30]

Another theory is that FRBs could be triggered by active galactic nuclei, or AGNs—superluminous regions at the centers of some large galaxies that are thought to be powered by a supermassive black hole lurking there. Many AGNs have jets that could generate FRBs, but there's a problem: AGNs usually don't exist in dwarf galaxies. As a result, while catching the repeater was an achievement, it also created new scientific problems: might there be two different types of FRBs after all? Those that repeat and those that don't?[31]

Shortly after the PALFA collaboration published their astonishing results on the host galaxy of the repeater, in 2017, Loeb sparked a media frenzy—by suggesting that FRBs could be of alien origin after all. He had been mulling over his concept for a light sail technology that might get people to Alpha Centauri—and decided to flip the concept around. What, he wondered, if the FRBs were simply the byproduct of solar-powered radio transmitters—mega interstellar light sails pushing giant spaceships across galaxies? The story certainly helped to make FRBs more of a household term, even though the vast majority of astronomers and astrophysicists did not agree with Loeb's theory.[32] "If we're detecting light sails from a dwarf galaxy at a gigaparsec, we should be detecting many, many more of them from galaxies that are much, much closer to us," says Chatterjee. "There's about as much chance that it's the Death Star being blown up in a galaxy far far away."[33]

Still, aliens or not, most astronomers have been very happy to work alongside the Breakthrough Listen initiative—which has helped them to upgrade their telescopes and obtain crucial scientific results. Detecting FRBs has quickly become one of the main objectives of Breakthrough Listen, and just months after the repeater was localized, the Breakthrough Listen team using the Green Bank Telescope threw more weight behind the magnetar idea. First, astronomers looking at Arecibo data of some of the repeating flashes discovered that whatever was producing them may be living in an extreme, highly magnetized environment. The magnetic field near the source seemed so strong that it was twisting its radio waves, an effect known as Faraday rotation. GBT confirmed the result. While scanning the skies for signs of aliens, the Breakthrough Listen researchers decided to point it at the repeater—and the telescope observed twenty-one additional bursts at even higher frequencies, all with the same highly twisted Faraday rotation.[34]

At Parkes, too, the influence of Breakthrough Listen is obvious. As mentioned earlier, I visited the telescope's control room in February 2019. To reach it, I went up a flight of stairs at the circular tower under the antenna, where every button and door nostalgically screams 1960s. The control room is now full of modern computers that astronomers use to remotely control the dish in order to observe pulsars. Another flight of stairs, and I was in the data storage room, packed with columns and columns of computer drives with blinking lights. An entire column of hard drives, one meter (3 ft) high and 2.7 meters (9 ft) wide, belongs to Breakthrough Listen—the core of a cutting-edge recording system that enables astronomers to search for every possible radio signal in twelve hours of data, surpassing what was previously possible. Bailes, who works on both FRB research and Breakthrough Listen, took a smiling selfie with me in front of Milner's drives.

Whatever creates the FRBs—magnetars, alien life, or anything else—identifying that source and catching more than just a few FRBs here and there clearly would require an infusion of new tech. Upgrading veteran single dishes was good, but not good enough: because an FRB is thought to go off every second somewhere in the Universe, to see them they need ways of looking at the entire sky at the same time. Single-dish telescopes like Parkes, GBT, and Arecibo have a fairly small field of view—they can point only at a specific area of the sky, which means that catching these erratic flashes very much depends on luck. So Milner turned to one of the two Square Kilometre Array (SKA) precursors, MeerKAT, making it part of Breakthrough Listen. Once the SKA is built, it will consist of about two thousand high- and mid-frequency antennas and aperture arrays as well as roughly one million low-frequency dishes.[35] The first SKA dish was assembled at the MeerKAT site in April 2019, shipped there from China; I watched engineers poke its pedestal's electronic innards, the dish resting nearby.

This South African array and a few other new generation telescopes are starting to revolutionize the nascent field of FRB research. And they are doing it full tilt: two other arrays have recently localized different one-off flashes for the first time. One of them is another SKA precursor, the Australian Square Kilometer Array Pathfinder (ASKAP) in Western Australia, and the other is Caltech's Deep Synoptic Array-10 at the Owens Valley Radio Observatory (OVRO) near Big Pine, California. In addition, a totally new telescope in Canada, the Canadian Hydrogen Intensity Mapping Experiment (CHIME), is now catching FRBs by the dozen. It won't be much longer, astronomers think, until they finally crack the mystery of the source of these bursts. So before it ceases being a mystery and becomes just as mundane as, say, the gamma-ray bursts, I head to the Okanagan Valley, a radio-quiet zone surrounded by the mountains of Canadian province of British Columbia, a region also famous for its wines.

CHIME: Detecting FRBs in Canadian Wine Country

It's sunny but so chilly that in desperation I try to pull my baseball cap deeper onto my head. It's early October 2019, and I'm in a fairly small open space framed by mountains on all sides. In front of me is a huge metallic structure composed of four open, hundred-meter (109-yard) long, U-shaped half-cylinders next to each other, each studded with thousands of antennas. Made of metal mesh, the telescope looks like a half-pipe park for skaters or snowboarders—albeit a huge one, with an area the size of five hockey rinks. This is CHIME.

Somehow, the telescope reminds me of Molonglo, a pulsar (and, more recently, FRB) detector in Australia. The two constructions are very, very different, but their sheer scale and unique shapes are so unusual that my brain automatically starts to compare them. CHIME is a novel digital radio telescope that works with low frequencies of 400 to 800 MHz and can scan extremely wide areas of the sky; as its many antennas detect radio waves, a central computer renders the composite image.[36] It's Canada's largest telescope, with a footprint larger than six NHL hockey rinks, comprised of a 100 meter by 100 meter collecting area.

I traveled here from London via Montreal, where I grew up, and where I visited my Canadian alma mater, McGill University. There, I spoke to the astronomer Vicky Kaspi who, as you read in Chapter 4, discovered a totally new way of timing magnetars by applying a known technique of timing radio pulsars to x-ray astronomy. More recently, she's been doing a lot of work on FRBs—in fact, Paul Scholz, who spotted the first repeating FRB from the Spitler Burst in 2015 in the archival data of the PALFA survey, was her graduate student. Kaspi herself was the principal investigator of PALFA. Two years before that discovery though, in 2013, Kaspi read that groundbreaking paper by Dan Thornton and his colleagues that detailed the discovery of four FRBs in the Parkes data—the first haul after Lorimer burst. It was this paper "that really made me believe FRBs were real," Kaspi says. The paper prompted her and her colleagues to come up with a cunning plan as soon as she learned what CHIME would be capable of.[37]

CHIME is part of the Dominion Radio Astrophysical Observatory (DRAO), a research facility dating from the 1960s and at some point involved in building the correlator for the Very Large Array in New Mexico—the VLA's brain, the key element that made it functional. Built by a consortium of cosmologists from McGill, the University of Toronto, UBC, and others, CHIME was placed at DRAO mainly because there was land available, the radio noise in the area is low, and there was infrastructure and expertise such as radio engineers on hand, says Kaspi.[38] The new telescope's original purpose was cosmological: to try to shed some light on the nature of dark energy by very accurately measuring the acceleration of the Universe's expansion. Because it can see so much of the sky at any one time, its other goals are to study our galaxy's magnetic fields and neutron stars, as well as to time them, in an effort to detect gravitational waves from colliding supermassive black holes.

"Once CHIME was being built, it became clear that it would be a great fast radio burst detector," Kaspi says, sitting in her office at McGill. "I wanted to learn as much as I could about it and understand it, because I wanted to write a proposal to get money so that we could also use it for fast radio bursts. And that was a fairly large proposal—it meant millions of dollars—so I had to learn a lot about it."[39]

Kaspi teamed up with several cosmologists, including Matt Dobbs, Keith Vanderline, Mark Halpern, Gary Hinshaw, and others. They put together a proposal, led by Kaspi, for the Canada Foundation for Innovation—asking for a grant to add FRB-related equipment to CHIME. It was a success—in 2015 they got C$5.6 million (which became part of the total cost of around $20 million for building CHIME).[40] But she couldn't just sit back and wait for CHIME to get the extra equipment: she wanted to help with building it, much like Jocelyn Bell who picked up a hammer and pounded wooden poles into the ground in order to speed along construction of her array. In the summer of 2017, Kaspi worked on the telescope along with dozens of other scientists, engineers, and students, attaching cables to different parts of the telescopes. The cables were laid by mainly students and postdocs, she says. "I spent my

Vicky Kaspi with her students at CHIME. (Courtesy of Vicky Kaspi)

career as an observer, where I use instruments that are built for me and for which generally there's a manual and an expert on call," Kaspi laughs. "And here, we had been laying hundreds of cables, some in ditches, some along the cylinders." Each cable had to be plugged in and torqued very specifically—not too loose because then it's leaky, and not too tight because then it's not good for the connector. "I was probably the only person there who had never heard of a torque wrench before!" chuckles Kaspi.[41]

The cabling of the telescope was an enormous task. They had to keep the birds and the weeds underneath the structure at bay, shoo away the neighboring farmer's cows, and make sure that the telescope didn't get damaged in the rain and the snow. And then there were the bears and rattlesnakes. Kaspi recalls seeing a snake a few yards from her, slithering away at a fast clip. "It was zipping down the road—I just was amazed at the physics of the locomotion, how it makes itself move, I'd never seen that," she says. While outside it was the blistering Okanagan summer, inside the huts where cabling also had to be done, it was rather chilly, because the electronics have to be cooled to a low temperature. The electronics were all custom-built, sophisticated circuit boards, developed by Kaspi's McGill faculty colleague Matt Dobbs. "It was a whole process, but I listened to a lot of music, it was actually quite relaxing and therapeutic," says Kaspi. "Just days and days of doing nothing but cabling, a lot of it together with a grad student Ziggy Pleunis. Our fingers were quite raw by the end of it. But it was worth it."[42]

Full of Kaspi's stories, I flew to Vancouver and then on a small, two-row propeller plane to the city of Kelowna in the Canadian province of British Columbia. Next, I drove for a couple of hours along the picturesque Okanagan Lake, passing vineyard after vineyard. The Okanagan Valley is famous for its unique microclimate—it's milder here in winter than elsewhere in Canada, and summers are brutally hot. The conditions allow the locals to grow apples and grapes in abundance and make fantastic wine, with plenty of wine-tasting places all along the famous Naramata Road. "We are the Hawaii of Canada," a cashier girl at a trendy coffee shop told me proudly. I stopped there on the way to CHIME, in the nearby town of Penticton, wedged between the Okanagan and Skaha lakes. The name "Penticton" loosely translates from the Okanagan dialect of the indigenous Salish tribe as "a place to stay forever," which also refers to the continuous flow of the Okanagan River out of the lake—and I did feel like time was passing more slowly here; the serenity is nearly palpable. Leaving the coffee shop behind, I drove by a small but well-maintained pet cemetery, in the woods just minutes away from the telescope. A few more turns, and the sign "Please switch off your phone" tells me very clearly that CHIME is nearby.

When I get to the detector, astronomer Tom Landecker meets me. He's a smiley, chatty Australian scientist and engineer who moved to Canada some three decades ago—and has been at DRAO ever since. As he shows me the cables that the CHIME team of students and postdocs had laid, he tells me that the telescope was built on the cheap, by a local company that typically builds supermarkets, with the same kind of materials. We go up the ladder to reach the top of the first half-cylinder and I look around. The beauty of this place is stunning, and I keep expecting a bear to walk out of the woods. The nature all around is so wild that it's as if it has gracefully allowed these few humans to squat on this tiny space with their radio ears to listen to the Universe.

As Kaspi and her collaborators were busy upgrading CHIME to turn it into an FRB machine, what she didn't know at the time was that just 20 km (12.4 miles) from DRAO, the family of a young astrophysicist named Ryan Shannon was waiting for him to come back home to British Columbia for Christmas that year. Back when Lorimer was mulling over the *Nature* rejection in 2007, Shannon was doing his PhD in physics at Cornell University in Ithaca, New York—sharing an office with none other than Laura Spitler.

Even though Shannon never visited DRAO as a kid, and although CHIME hadn't been built yet when he left, home, he thinks that being so close to it may have somehow influenced his decision to study radio astronomy. He visited the telescope for the first time on January 2, 2019. A few days earlier, he had been snowshoeing in a picturesque British Columbian forest not far from his parents' house when he unexpectedly bumped into a researcher he knew from Australia who was now a postdoc at CHIME. "Small world!" chuckles Shannon, a young blond researcher, when I meet him at Swinburne University of Technology in Melbourne, just before joining Matthew Bailes on our road trip to Parkes.[43] Shannon's mother had wanted him to stay close to home, he says, but after his graduation from Cornell in 2011, he moved to Australia to work on pulsars.

ASKAP: Next-Gen Dish Cluster in Western Australia

Ryan Shannon got pulled into research on fast radio bursts only gradually, seeing it, like many pulsar astronomers, as an unfinished chapter in a book on neutron stars—a pithy description once given to me by his Cornell colleague Shami Chatterjee. And just like that, Shannon and other researchers from CSIRO and Curtin University in Australia realized that they also had a perfect FRB machine being built right in their own backyard: the Australian Square Kilometer Array Pathfinder (ASKAP).[44]

Similar to CHIME and MeerKAT, ASKAP wasn't originally meant for FRBs. It was conceived before Duncan Lorimer discovered the very first burst, and construction started in 2010. ASKAP's main aims have always been to study galaxy formation using extragalactic surveys of hydrogen gas, to investigate the evolution of galaxies as well as galactic magnetic fields and—crucially—to discover more about radio transient sources, which are not just pulsars but also any other unexpected radio flashes flickering in the sky, such as FRBs.[45] "Unfortunately, we missed nearly all FRBs in the first decade of realizing these things exist," says Shannon.[46]

As more and more bursts were detected though, in around 2015 Shannon and his colleagues at CRAFT collaboration, which stands for Commensal Real-time ASKAP Fast Transients, had an idea. It occurred to them that you didn't need a gigantic dish like Parkes to search for the enigmatic ultra-brief radio flashes; instead, what you need most is a telescope with a wider field of view. This is how a fast radio burst detection system for ASKAP was born, and CRAFT "has been the focal point of these efforts," says Shannon's colleague at Swinburne, astrophysicist Adam Deller. "The CRAFT collaboration realized that if bright FRBs are relatively numerous, then the incredibly wide field of view ASKAP offered meant that a good sample of FRBs could be found even while the incomplete telescope was still being commissioned." So CRAFT, with members spanning Australian and international universities and the CSIRO, set to the task of building a fast radio burst detection system for the array.

With its thirty-six antennas each twelve meters (39 ft) across, the array was finally completed in 2019—but the first ASKAP FRBs were found well before then.[47]

Not many astronomers visit ASKAP—it's been built to be operated fully remotely. A bunch of dishes in a mostly unpopulated area in the Murchison Shire of Western Australia, the array is a curious sight for the small Indigenous population—the Wajarri Yamatji people—who are the original owners of the land. Shannon has been to the site as part of an outreach effort, explaining the purpose of the telescope to the locals and the research it is used for. "I think they're very, very happy for us to be there," he says, adding that there haven't been any protests against the telescope like there were, say, in Hawaii in the run-up to the planned construction of a telescope on sacred Mount Kau. "It is obviously easier to explain optical astronomy to people than radio astronomy, because that's something you can see with your eyes," adds Shannon.[48] But people were still very engaged and interested, he says. Each of the thirty-six antennas has been named by the locals, with names like Bundarra (stars), Wilara (the Moon), and Jirdilungu (the Milky Way).

To get there, Shannon had to fly to Perth, and then on a smaller plane to Murchison. From there, he flew on a tiny single propeller plane (to avoid driving for five hours across 150 km, or 93 miles, of dirt roads) to the site. Driving is especially not recommended in the rain, he laughs, because roads turn to mud.[49]

While MeerKAT and ASKAP are both SKA precursors and both can hunt for FRBs, technologically they are quite different—and thus are suited to finding FRBs with different characteristics. They both look at the southern sky, which is a better vantage point for seeing the Milky Way's Galactic Center than is the northern half of the globe, complementing older telescopes like Parkes and Arecibo. MeerKAT's dishes have much more sensitive receivers, able to detect objects that are farther away, while ASKAP is less sensitive but with a much wider field of view, meaning that astronomers can observe a much larger portion of the sky, and so can study closer objects, and their surroundings, in much more detail.

Each ASKAP antenna is equipped with a system called a phased-array feed, which is similar in some ways to the technology that allows you to take images in your camera. Most single-dish radio telescopes traditionally had the equivalent of only one pixel, although Parkes with its multi-beam receiver had thirteen pixels. All ASKAP dishes working together form a virtual telescope with thirty-six pixels—quite an improvement for their field of view. To net FRBs, the dishes can be made to point in different directions, similar to how a fly's eye works. And while ASKAP was still in the commissioning phase, in a configuration with only ten dishes, the astronomers caught twenty FRBs.[50] That's pretty good, laughs Shannon, "considering in the previous decade, people [got] only twenty-seven."[51]

Besides netting the bursts, ASKAP has also made its mark by employing another advanced technique to find the source of the burst, even if it happens only once. To achieve this, the team points all the dishes in the same direction, so they all look at the same portion of sky—and wait for an FRB to go off. It's a massive technical undertaking to make all thirty-six antennas work in sync. But this way, the telescope becomes a radio interferometer similar to MeerKAT and can make very precise images of specific objects in space. Once an FRB does flash, the idea is to look at the sky for that millisecond where the burst has come from and make an image of the location during that millisecond, in real time.[52] "What you will see then is only the burst and nothing else because the burst outshines everything else," says Shannon.[53] But it's not possible to search for FRBs using interferometry constantly—the computational load would be far too large. So, Deller chimes in, the neat trick the researchers also play is to look for the FRBs the same way as the fly's eye search—but after adding the signals from the different dishes together to improve the sensitivity, since they are now all looking at the same patch of sky. Then, when ASKAP detects a signal, within a second or so a download of the raw data gets triggered from the telescopes to perform the interferometry and imaging offline, which is slower than real time.

On the morning of September 25, 2018, Shannon and his colleague Adam Deller were looking at the results of the searches from the night before. Excited, they saw that they had successfully detected a fast radio burst with all the antennas. The next step was to attempt to pinpoint its location. Deller took the original raw data, played it back, and made the image. It was technically very challenging, says Deller, because "compared to just noticing that an FRB had occurred, making an image to see where it came from needs a lot of extra calibration." The signals from all dishes had to be precisely aligned in time—to the sub-nanosecond level—alongside the digital signal processing to produce the image. It took Deller two days to gather all the data and line them up through a process called correlation, but finally he managed to create an image on the computer screen: a map showing where the FRB had gone off. "And we were like, oh my god, there's a galaxy there! It was really, really exciting," smiles Deller.[54]

Surprisingly, the team found that the home galaxy of the burst, some four billion light-years away, is very different from the one where the repeater lives. "This galaxy is bigger and it has less star formation," says Shannon. "It has a lot more just old stars." Also, while the repeater seems to be in a very dense and fairly highly magnetized plasma, consistent with the magnetar theory that imagines magnetars embedded in dense and magnetized supernova remnants, the environment of the one-off localized burst is nothing like that. We also know that the first repeater was embedded in a very bright radio source, given how far away the burst was. "With the burst that we've localized, we see nothing like that there," says Shannon. "So are these things even related to one another?"[55]

During the same week in August 2019 that Shannon's paper on the localization of a one-off burst was published, another paper came out—by a team at Caltech's Owens Valley Radio Observatory (OVRO) in California's Sierra Nevada mountains. Astronomers there also traced an FRB back to its source, a galaxy 7.9 billion light-years away, using ten 4.5-meter (14.8-ft) antennas forming the Deep Synoptic Array-10, a precursor to the future Deep Synoptic Array with 110 radio dishes. This one-off burst also lives in a galaxy with very little star formation.[56]

Shannon thinks that what triggered these two bursts may have something to do with evolved stars, such as merging white dwarfs or merging neutron stars. Maybe it will turn out that the repeating sources are all generated by young magnetars and the non-repeating ones are created by mergers or something else. How can something make such a luminous radio burst, many orders of magnitude outside of anything we've seen from within our own Milky Way? "Pulsars are amazing. And these things are a billion times more amazing," Shannon says, excitedly.[57]

Back in Shannon's home country of Canada, CHIME is now catching FRBs by the dozen. In January 2019, the CHIME team reported the detection of six repeat bursts from FRB 180814.J0422+73—the second "repeater" after the Spitler Burst and one of the thirteen FRBs detected by CHIME during its pre-commissioning phase in July and August 2018.[58] Then later that same year, the Canadian instrument year, it netted seventeen more—eight reported in August and nine more in January 2020.[59] Astronomers localized the second repeater in October 2019—it lives in a galaxy quite similar to our own Milky Way, one starkly different to the home of the first repeating burst. As this book goes to print, there will certainly be many more FRB discoveries made by CHIME, ASKAP, Parkes, and other FRB-hunting machines around the globe, because now they know much better what they are looking for.

Slowly but surely, this unfinished chapter on the mysteries of fast radio bursts seems to be approaching the finish line. As for the SETI theory, Lorimer says that Loeb's reasoning around his alien-related FRB origin models is not fundamentally wrong—the energetics based on the observations are consistent. He personally prefers to find the simplest explanation for natural phenomena in space, but he thinks that until we rule everything out with observations, scientifically sound theories should stand.[60] A definitive answer may be coming soon. Astronomers are certain that they will find the origin of these mesmerizing radio bursts in the sky in the very near future—and so far, most bets are on neutron stars.

Epilogue

"WE'RE MADE OF STAR STUFF."

That's what Carl Sagan once famously said, and as I learned during these past few years, quite a bit of our world is made of neutron star stuff as well. After all, it takes the collision of two neutron stars to unleash the forces needed to create most of the very heavy elements, such as silver, gold, platinum, and many others.

Detecting the gravitational waves from the faraway collision of two neutron stars was one of the most important scientific discoveries of the past decade. It united researchers from many different fields of physics and became a turning point for a new approach to astrophysics: multi-messenger astronomy. Measuring and understanding gravitational waves, and being able to locate and observe distant and long-gone cataclysmic events millions of light-years away as a result, are much more important than just detecting ripples in spacetime and proving Einstein right. These breakthroughs have given scientists an additional sense with which to perceive the Universe—and opened up an exciting, vibrant field of research that may help us to understand the nature of many still-puzzling phenomena. And scientists are finally getting close to being able to create the tools they need to probe the guts of neutron stars, which—despite decades of research—are still one of the most perplexing mysteries of the cosmos. Understanding neutron stars will not only allow us to figure out what makes these ultra-dense objects tick, but will also give us insights into the fundamental workings of the Universe.

It was this rapidly unfolding mystery of what neutron stars are and the role they play in the cosmos that took me around the world. It also took me to the edge of physics, the very boundary of where we humans can grasp how our Universe ticks. This book is my tribute to all the researchers and thinkers, the astronomers, cosmologists, astrophysicists, nuclear and particle physicists, and gravitational wave researchers who study all the different, enigmatic aspects of neutron stars. Those who tirelessly, year after year, take measurements of the arrival time of radio pulses from pulsars scattered across the sky. Their endless calculations and technological breakthroughs have allowed humankind to measure extremely accurately the mass of these faraway objects and the distance to them—a mind-blowing achievement. It's also a tribute to the engineers who are building the next generation of cutting-edge telescopes and detectors, to investigate the entire electromagnetic and astroparticle spectrum. These astonishing machines include the Square Kilometer Array (SKA); the Facility for Antiproton and Ion Research (FAIR), which is the next collider to probe the conditions in the inner core of a neutron star; and the Laser Interferometer Space Antenna, (LISA), the space detector for low frequency gravitational waves set to be launched in 2034 and which, along with pulsar timing arrays, may tell us whether supermassive black holes in the center of galaxies really do merge. Over the next decade or so, as they go online, these instruments will radically transform the way we perceive the cosmos.

Our Universe is littered with dead stars that have run out of fuel and died a quiet death to become undetectable brown dwarves, or that have disappeared with one spectacular final explosion, leaving behind nothing but a black hole immersed in hot clouds of gas and dust. But then there are the cosmic zombies—the neutron stars. Ultra-dense, a smidgen or two away from turning into a black hole, they have a life beyond death: they send out radio waves, gamma rays, x-rays, and maybe the enigmatic fast radio bursts. Some neutron stars have become unobservable by now, lurking unseen and undetectable in the depths of space; others are spinning at unimaginable speeds, powered by the steady flow of matter from their faithful binary companions.

This scientific journey of discovery is not over, not by a long shot. By unlocking more and more mysteries of neutron stars, we will keep on learning some of the deepest secrets of our Universe. And there are plenty more to discover.

Notes

Prologue

1. M. Branchesi, phone interview with author, Oct. 19, 2018.

1 · A Collision That Shook the Cosmos

1. M. Branchesi, phone interview with author, Oct. 19, 2018.
2. B. P. Abbott et al. (LIGO Scientific Collaboration and Virgo Collaboration), "GWTC-1: A Gravitational-Wave Transient Catalog of Compact Binary Mergers Observed by LIGO and Virgo during the First and Second Observing Runs," *Physical Review X* 9, no. 3 (Sept. 4, 2019): 031040.
3. Details about Branchesi's experiences during these first days are all from Branchesi, interview, Oct. 19, 2018.
4. B. P. Abbott et al. (LIGO Scientific Collaboration and Virgo Collaboration), "GW170817: Observation of Gravitational Waves from a Binary Neutron Star Inspiral," *Physical Review Letters* 119 (Oct. 16, 2017): 161101.
5. E. Hamilton, "What Is Multi-Messenger Astronomy?" UW News, University of Wisconsin, Madison, n.d., https://news.wisc.edu/what-is-multi-messenger-astronomy.
6. "LIGO Congratulates IceCube on Multimessenger Astronomy Success," news release, LIGO Laser Interferometer Gravitational Wave Observatory, July 16, 2018, https://www.ligo.caltech.edu/WA/news/ligo20180716.

7. B. P. Abbott et al. (LIGO Scientific Collaboration and Virgo Collaboration), "Multi-Messenger Observations of a Binary Neutron Star Merger," *Astrophysical Journal Letters* 848, no. 2 (Oct. 16, 2017): L12.
8. J. Kluger, "Marica Branchesi," *Time* 100: The Most Influential People of 2018, April 30–May 7, 2018, https://time.com/collection/most-influential-people-2018/5238152/marica-branchesi.
9. "Neutron Stars," Imagine the Universe! NASA, Goddard Space Flight Center, Mar. 2017, https://imagine.gsfc.nasa.gov/science/objects/neutron_stars1.html.
10. R. N. Manchester, "Millisecond Pulsars, Their Evolution and Applications," *Journal of Astrophysics and Astronomy* 38 (Sept. 2017): 42.
11. C. Skelly, "NASA Continues to Study Pulsars, 50 Years after Their Chance Discovery," NASA, Aug. 1, 2017, https://www.nasa.gov/feature/goddard/2017/nasa-continues-to-study-pulsars-50-years-after-their-chance-discovery.
12. Abbott et al., "GW170817."
13. "Astronomers See Light Show Associated with Gravitational Waves," news release 2017-30, Harvard and Smithsonian Center for Astrophysics, Oct. 16, 2017, https://www.cfa.harvard.edu/news/2017-30.
14. S. Perkins, "Neutron Star Mergers May Create Much of the Universe's Gold," *Science News,* Mar. 20, 2018, https://www.sciencemag.org/news/2018/03/neutron-star-mergers-may-create-much-universe-s-gold.
15. M. Branchesi, email interview with author, Sept. 15, 2019.
16. "A Brief History of Isaac Newton's Apple Tree," Department of Physics, University of York, n.d., https://www.york.ac.uk/physics/about/newtonsappletree.
17. T. Davis, D. Rickles, and S. Scott, "Understanding Gravity—Warps and Ripples in Space and Time," Australian Academy of Science, n.d., https://www.science.org.au/curious/space-time/gravity.
18. "The Binary Pulsar PSR 1913 + 16," course website for Astronomy 201, Department of Astronomy, Cornell University, n.d., http://hosting.astro.cornell.edu/academics/courses/astro201/psr1913.htm.
19. "Gravity Investigated with a Binary Pulsar," press release, Royal Swedish Academy of Sciences, Oct. 13, 1993, https://www.nobelprize.org/prizes/physics/1993/press-release.
20. "Timeline," LIGO website, n.d., https://www.ligo.caltech.edu/page/timeline.
21. H. Johnston, "Virgo Bags Its First Gravitational Waves," Physicsworld.com, Sept. 27, 2017, https://physicsworld.com/a/virgo-bags-its-first-gravitational-waves.
22. V. Baschi, in-person interview with author, Sept. 30, 2019, European Gravitational Observatory, Santo Stefano a Macerata, Cascina, Italy.
23. S. Katsanevas, in-person interview with author, Sept. 30, 2019, European Gravitational Observatory, Santo Stefano a Macerata, Cascina, Italy.

24. Ibid.
25. "Virgo in a Nutshell," Virgo website, n.d., http://public.virgo-gw.eu/virgo-in-a-nutshell.
26. "About the Instruments and Collaborations," Gravitational Wave Open Science Center, n.d., https://www.gw-openscience.org/links.
27. Branchesi, interview, Oct. 19, 2018.
28. Ibid.
29. B. Metzger, email to author, Nov. 6, 2018.
30. Branchesi, interview, Oct. 19, 2018.
31. S. Smartt, phone interview with author, May 23, 2017.
32. "Memorandum of Understanding between VIRGO on One Side and the Laser Interferometer Gravitational Wave Observatory (LIGO) on the Other Side," Mar. 20, 2014, https://dcc.ligo.org/public/0001/M060038/002/Main_MOU_signed.pdf.
33. Branchesi, interview, Oct. 19, 2018.
34. J. Chu, "LIGO and Virgo Make First Detection of Gravitational Waves Produced by Colliding Neutron Stars," MIT News Office, Oct. 16, 2017, http://news.mit.edu/2017/ligo-virgo-first-detection-gravitational-waves-colliding-neutron-stars-1016.
35. Branchesi, interview, Oct. 19, 2018.
36. S. Nissanke, phone interview with author, Oct. 22, 2018.
37. Abbott et al., "GW170817."
38. C. Messick, email interview with author, Nov. 21, 2018.
39. Ibid.
40. Branchesi, interview, Oct. 19, 2018.
41. Messick, interview, Nov. 21, 2018.
42. R. C. Essick, "LIGO/Virgo G298048: Fermi GBM Trigger 524666471/170817529: LIGO/Virgo Identification of a Possible Gravitational-Wave Counterpart," GCN Circular, Aug. 17, 2017, https://gcn.gsfc.nasa.gov/other/G298048.gcn3.
43. "First Ever Optical Photons from a Gravitational Wave Source," GW170817/SSS17a, UCSC News, n.d., https://ziggy.ucolick.org/sss17a.
44. "Team," GRAWITA, Gravitational Waves at INAF, n.d., https://www.grawita.inaf.it/team-2.
45. Details in this section about when Berger received the message and his response are from E. Berger, phone interview with author, Sept. 28, 2017.
46. K. Alexander, phone interview with author, Oct. 10, 2018.
47. Ibid.
48. Berger, interview, Sept. 28, 2017.
49. Branchesi, interview, Sept. 15, 2019.

50. C. Day, "Bursts from the Cold War," *Physics Today*, Aug. 6, 2013, https://physicstoday.scitation.org/do/10.1063/PT.5.010233/full.
51. Alexander, interview, Oct. 10, 2018.
52. Chu, "LIGO and Virgo Make First Detection."
53. Branchesi, interview, Oct. 19, 2018.
54. Ibid.
55. B. Metzger, email interview with author, Sept. 24, 2019.

2 · Discovering Neutron Stars . . . and Little Green Men?

1. M. Longair, in-person interview with author, Jan. 11, 2019, University of Cambridge, Cambridge, UK.
2. J. Bell Burnell, "Pliers, Pulsars, and Extreme Physics," *Astronomy & Geophysics* 45, no. 1 (Feb. 1, 2004): 1.7–1.11.
3. S. Devons, "Rutherford Laboratory," Department of Physics, The Cavendish Laboratory, University of Cambridge, n.d., https://www.phy.cam.ac.uk/history/years/rutherford.
4. "1963: Maarten Schmidt Discovers Quasars," Carnegie Institution for Science, n.d., https://cosmology.carnegiescience.edu/timeline/1963.
5. A. G. Levine, "Holmdel Horn Antenna: Holmdel, New Jersey. The Large Horn Antenna and the Discovery of Cosmic Microwave Background Radiation," *APS Physics* (2009), https://www.aps.org/programs/outreach/history/historicsites/penziaswilson.cfm.
6. L. Badash, "Ernest Rutherford, British Physicist," Encyclopedia Britannica, Aug. 26, 2019, https://www.britannica.com/biography/Ernest-Rutherford/McGill-University; Longair, interview, Jan. 11, 2019.
7. "James Chadwick, Biographical," The Nobel Prize, Dec. 9, 2019, https://www.nobelprize.org/prizes/physics/1935/chadwick/biographical.
8. Longair, interview, Jan. 11, 2019.
9. A. S. Burrows, "Baade and Zwicky: 'Super-novae,' Neutron Stars, and Cosmic Rays," *PNAS* 112, no. 5 (Feb. 3, 2015): 1241–1242.
10. W. Clavin, "Zwicky Transient Facility Opens Its Eyes to the Volatile Cosmos," California Institute of Technology, Nov. 14, 2017, https://www.caltech.edu/about/news/zwicky-transient-facility-opens-its-eyes-volatile-cosmos-80369.
11. Longair, interview, Jan. 11, 2019.
12. Burrows, "Baade and Zwicky."
13. J. R. Oppenheimer and G. M. Volkoff, "On Massive Neutron Cores," *Physical Review* 55, no. 4 (1939): 374–381.
14. Longair, interview, Jan. 11, 2019.

15. V. Trimble, "Oppenheimer, J. Robert," *Biographical Encyclopedia of Astronomers*, 2007, https://link.springer.com/referenceworkentry/10.1007%2F978-0-387-30400-7_1037.
16. Longair, interview, Jan. 11, 2019.
17. J. Sarkissian, "Parkes and 3C273: The Identification of the First Quasar," CSIRO, n.d., https://www.parkes.atnf.csiro.au/people/sar049/3C273.
18. "1963: Maarten Schmidt Discovers Quasars"; Longair, interview, Jan. 11, 2019.
19. Longair, interview, Jan. 11, 2019.
20. Ibid.
21. J. L. Linsky, B. J. Rickett, and S. Redfield, "The Origin of Radio Scintillation in the Local Interstellar Medium," *Astrophysical Journal* 675, no. 1 (2008).
22. Burnell, "Pliers, Pulsars, and Extreme Physics."
23. M. Longair, "A Brief History of Radio Astronomy in Cambridge," Cavendish Astrophysics, University of Cambridge, 2016, https://www.astro.phy.cam.ac.uk/about/history.
24. Longair interview, Jan. 11, 2019.
25. J. Bell Burnell, in-person interview with author, Mar. 1, 2019, Oxford University, Oxford, UK.
26. Burnell, "Pliers, Pulsars, and Extreme Physics."
27. Longair interview, Jan. 11, 2019.
28. Ibid.
29. Burnell, "Pliers, Pulsars, and Extreme Physics."
30. Burnell, interview, Mar. 1, 2019.
31. K. Kellermann and B. Sheets, "Serendipitous Discoveries in Radio Astronomy," *Proceedings of a Workshop Held at the National Radio Astronomy Observatory Green Bank, West Virginia* (May 4–6, 1983): 160–170, http://library.nrao.edu/public/collection/02000000000280.pdf.
32. Burnell, interview, Mar. 1, 2019.
33. Ibid.
34. Ibid.
35. Kellermann and Sheets, "Serendipitous Discoveries."
36. Longair, interview, Jan. 11, 2019.
37. Burnell, "Pliers, Pulsars, and Extreme Physics."
38. Kellermann and Sheets, "Serendipitous Discoveries."
39. Burnell, interview, Mar. 1, 2019.
40. Longair, interview, Jan. 11, 2019.
41. Burnell, interview, Mar. 1, 2019.
42. Kellermann and Sheets, "Serendipitous Discoveries."

43. Ibid.
44. Burnell, interview, Mar. 1, 2019.
45. Longair, interview, Jan. 11, 2019.
46. A. Hewish et al., "Observation of a Rapidly Pulsating Radio Source," *Nature* 217, no. 5130 (Feb. 1968).
47. "Anthony Michaelis," *Telegraph,* Mar. 28, 2008, https://www.telegraph.co.uk/news/obituaries/1583056/Anthony-Michaelis.html.
48. T. Gold, "The Origin of Cosmic Radio Noise," *Proceedings of Conference on Dynamics of Ionized Media*, University College, London, 1951; F. Pacini, "Energy Emission from a Neutron Star," *Nature* 216 (1967): 567–568, https://www.nature.com/articles/216567a0.
49. T. Gold, *Taking the Back off the Watch*, ed. Simon Mitton (Berlin: Springer-Verlag, 2012), 139.
50. "Nature's Astronomical Highlights," *Nature Astronomy* (Jan. 2, 2017), https://www.nature.com/collections/fmnhltzzlj/pulsars.
51. M. Bailes, in-person interview with author, Feb. 8, 2019, Parkes Telescope, Parkes, Australia.
52. Ibid.
53. J. Sarkissian, in-person interview with author, Feb. 9, 2019, Parkes Telescope, Parkes, Australia.
54. Ibid.
55. R. N. Manchester, "Pulsars at Parkes," CSIRO Astronomy and Space Science, Australia Telescope National Facility, Nov. 4, 2011, https://www.atnf.csiro.au/research/conferences/Parkes50th/ProcPapers/manchester.pdf.
56. "About Parkes Radio Telescope," CSIRO, Apr. 26, 2019, https://www.csiro.au/en/Research/Facilities/ATNF/Parkes-radio-telescope/About-Parkes.
57. Bailes, interview, Feb. 8, 2019.
58. All of the quotations by Cordes in this section are from J. Cordes, phone interview with author, Sept. 12, 2019.
59. Tim O'Brien, "Observations of Pulsars," course website for Frontiers of Modern Astronomy, Jodrell Bank Observatory, University of Manchester, n.d., http://www.jb.man.ac.uk/distance/frontiers/pulsars/section4.html.
60. "The Voyage to Interstellar Space," NASA, Mar. 27, 2019, https://www.nasa.gov/feature/goddard/2019/the-voyage-to-interstellar-space.

3 · When Stars Go Boom

1. "San Pedro de Atacama," Chile Travel, n.d., https://chile.travel/en/where-to-go/north-and-the-atacama-desert/san-pedro-atacama.

2. A. Azua-Bustos, "Unprecedented Rains Decimate Surface Microbial Communities in the Hyperarid Core of the Atacama Desert," *Nature Scientific Reports* 8, no. 16706 (Nov. 12, 2018).
3. "Very Large Telescope: The World's Most Advanced Visible-Light Astronomical Observatory," European Southern Observatory, n.d., https://www.eso.org/public/unitedkingdom/teles-instr/paranal-observatory/vlt.
4. "ALMA: In Search of Our Cosmic Origins," European Southern Observatory, n.d., https://www.eso.org/public/unitedkingdom/teles-instr/alma.
5. A. Ho, phone and email interviews with author, Apr. 4, 2019.
6. Sir Arthur Stanley Eddington, The Internal Constitution of the Stars (Cambridge, UK: Cambridge University Press, 1926); R. H. Fowler, "On Dense Matter," Monthly Notices of the Royal Astronomical Society 87, no. 2 (Dec. 10, 1926), https://academic.oup.com/mnras/article/87/2/114/1058897.
7. K. Thorne, *Black Holes and Time Warps: Einstein's Outrageous Legacy* (New York: Norton, 1995).
8. "Discovers Neutron, Embryonic Matter," *New York Times*, Feb. 28, 1932, https://www.nytimes.com/1932/02/28/archives/discovers-neutron-embryonic-matter-dr-james-chadwick-describes-it.html.
9. W. Baade and F. Zwicky, "Cosmic Rays from Super-Novae," *PNAS* 20, no. 5 (May 1, 1934): 259–263.
10. A. S. Burrows, "Baade and Zwicky: 'Super-novae,' Neutron Stars, and Cosmic Rays," *PNAS* 112, no. 5 (Feb. 3, 2015): 1241–1242.
11. D. G. Yakovlev et al., "Lev Landau and the Concept of Neutron Stars," *Uspekhi Fizicheskikh Nauk and P N Lebedev Physics Institute of the Russian Academy of Sciences* 56, no. 3 (2013).
12. "Oppenheimer–Volkoff limit," Oxford Reference website, n.d., https://www.oxfordreference.com/view/10.1093/oi/authority.20110810105528460.
13. H. T. Cromartie et al., "Relativistic Shapiro Delay Measurements of an Extremely Massive Millisecond Pulsar," *Nature Astronomy* (2019).
14. R. Ekers, in person interview with author, Feb. 12, 2019, Sydney.
15. J. de Swart, "Deciphering Dark Matter: The Remarkable Life of Fritz Zwicky," *Nature*, Sept. 3, 2019, https://www.nature.com/articles/d41586-019-02603-7.
16. Ekers, interview, Feb. 12, 2019.
17 A. Vaughan, email interview with author, Jan. 12, 2020.
18. A. Vaughan, phone interview with author, Apr. 30, 2019.
19. Ibid.
20. "A Surprise from the Pulsar in the Crab Nebula," news release, European Southern Observatory, Nov. 20, 1995, https://www.eso.org/public/news/eso9532.

21. "How Many Stars Are There in the Universe?" European Space Agency, n.d., https://www.esa.int/Our_Activities/Space_Science/Herschel/How_many_stars_are_there_in_the_Universe.
22. M. Cantiello, email interview with author, May 19, 2017.
23. "The Dawn of a New Era for Supernova 1987A," Hubblesite, NASA, Feb. 24, 2017, https://hubblesite.org/contents/news-releases/2017/news-2017-08.html.
24. S. Smartt, phone interview with author, May 23, 2017.
25. "The Dawn of a New Era for Supernova 1987A."
26. O. Yaron, phone interview with author, May 4, 2017, and email interview with author, Jan. 13, 2020.
27. O. Yaron et al., "Confined Dense Circumstellar Material Surrounding a Regular Type II Supernova," *Nature Physics* 13 (Feb. 13, 2017): 510–517.
28. Yaron, interview, May 4, 2017.
29. Ho, interviews, Apr. 4, 2019.
30. Yaron, interview, May 4, 2017.
31. Ho, interviews, Apr. 4, 2019.
32. Ibid.
33. A. Morris, "Birth of a Black Hole or Neutron Star Captured for First Time," news release, Northwestern University, Jan. 10, 2019, https://news.northwestern.edu/stories/2019/01/birth-of-a-black-hole-or-neutron-star-captured-for-first-time.
34. Ho, interviews, Apr. 4, 2019.
35. All of the quotations by Janka from this point until the end of the chapter are from T. Janka, email interview with author, Apr. 11, 2019.
36. "Stellar Evolution—The Birth, Life, and Death of a Star," NASA, Sept. 4, 2003, https://www.nasa.gov/audience/forstudents/9-12/features/stellar_evol_feat_912.html.

4 · Zombies and Starquakes

1. ASTRON (Netherlands Institute for Radio Astronomy), "LOFAR," n.d., https://astron.nl/telescopes/lofar.
2. S. M. Tan, email interview with author, Sept. 24, 2019.
3. "Restoration Dwingeloo Radio Telescope Kicks Off," news release, ASTRON (Netherlands Institute for Radio Astronomy), Apr. 27, 2012, https://www.astron.nl/news-and-events/news/restoration-dwingeloo-radio-telescope-kicks.
4. ASTRON, "LOFAR."
5. Tan interview, Sept. 24, 2019.
6. ASTRON, "LOFAR."
7. P. Ghosh, *Rotation and Accretion Powered Pulsars,* vol. 10 (Hackensack, NJ: World Scientific, 2007).

8. W. Becker and G. Pavlov, "Pulsars and Isolated Neutron Stars," ArXiv pre-print service, Aug. 19, 2002, https://arxiv.org/pdf/astro-ph/0208356.pdf.
9. NRAO (National Radio Astronomy Observatory), "Pulsar Properties," n.d., https://www.cv.nrao.edu/course/astr534/Pulsars.html.
10. J. Hessels, phone interview with author, May 10, 2019.
11. Ibid.
12. R. N. Manchester, "Millisecond Pulsars, Their Evolution and Applications," *Journal of Astrophysics and Astronomy* 38 (Sept. 2017): 42.
13. M. Matsuoka and K. Asai, "Simplified Picture of Low-Mass X-Ray Binaries Based on Data from Aquila X-1 and 4U 1608–52," *Publications of the Astronomical Society of Japan* 65, no. 2 (Apr. 25, 2013): 26.
14. D. Buckley, in-person interview with author, Apr. 20, 2019, Cape Town, South Africa.
15. Matsuoka and Asai, "Simplified Picture."
16. F. Reddy, "With a Deadly Embrace, 'Spidery' Pulsars Consume Their Mates," Goddard Space Flight Center, NASA, Aug. 7, 2017, https://www.nasa.gov/content/goddard/with-a-deadly-embrace-spidery-pulsars-consume-their-mates.
17. A. Archibald, "The End of Accretion: The X-Ray Binary / Millisecond Pulsar Transition Object PSR J1023 + 0038," *APS Physics* (Apr. 2015).
18. "Unique Double Pulsar Tests Einstein's Theory," news release, Jodrell Bank Centre for Astrophysics, The University of Manchester, n.d., http://www.jb.man.ac.uk/doublepulsar/news/press3.html.
19. C. Kouveliotou, phone interview with author, July 22, 2019.
20. Ibid.
21. C. Kouveliotou, R. C. Duncan, and C. Thompson, "Magnetars," *Scientific American* 288, no. 2 (Feb. 2003).
22. Kouveliotou, interview, July 22, 2019.
23. Ibid.
24. C. Thompson, phone interview with author, July 23, 2019.
25. Ibid.
26. Ibid.
27. Kouveliotou, interview, July 22, 2019.
28. Ibid.
29. Thompson, interview, July 23, 2019.
30. V. Kaspi, interview with author, Oct. 7, 2019, McGill University, Montreal.
31. Ibid.
32. V. Kaspi, email interview with author, Sept. 26, 2019.
33. Ibid.

34. F. P. Gavriil, V. M. Kaspi, and P. M. Woods, "Magnetar-like X-ray Bursts from an Anomalous X-ray Pulsar," *Nature* 419 (2002): 142–144.
35. V. M. Kaspi et al, "A Major Soft Gamma Repeater-like Outburst and Rotation Glitch in the No-Longer-so-Anomalous X-ray Pulsar 1E 2259 + 586," *Astrophysical Journal Letters* 588, no. 2 (Apr. 11, 2003).
36. Kaspi, interview, Oct. 7, 2019.
37. Kaspi, interview, Sept. 26, 2019.
38. Thompson, interview, July 23, 2019.
39. F. Camilo, in-person interview with author, Apr. 12, 2019, Cape Town.
40. Ibid.
41. Ibid.
42. Ibid.
43. Ibid.
44. Ibid.
45. Ibid.
46. Ibid.
47. F. Camilo, phone interview with author, July 31, 2019.
48. Thompson, interview, July 23, 2019.
49. Lyne, in-person interview with author, July 12, 2019, Jodrell Bank Observatory, UK.
50. Ibid.
51. M. Bailes, email interview with author, Mar. 12, 2020.
52. "Multibeam Receiver Description," CSIRO, n.d., https://www.atnf.csiro.au/research/multibeam/instrument/description.html.
53. D. A. Swartz et al., "The Ultraluminous X-ray Source Population from the Chandra Archive of Galaxies," *Astrophysical Journal* suppl. ser. 154, no. 2 (2004).
54. All the quotations in this "Deeper Dive" are from L. Townsend, in-person interview with author, Apr. 20, 2019, Cape Town, South Africa.
55. "Doppler Shift," Imagine the Universe! NASA, Goddard Space Flight Center, May 5, 2016, https://imagine.gsfc.nasa.gov/features/yba/M31_velocity/spectrum/doppler_more.html.
56. American Institute of Physics online history archive, "World Fame I," n.d., https://history.aip.org/history/exhibits/einstein/fame1.htm.

5 · Journey to the Center of a Neutron Star

1. R. N. Manchester, in-person interview with author, Feb. 12, 2019, CSIRO, Canberra.
2. "Polarization Light Waves and Color—Lesson 1—How Do We Know Light Is a Wave? Polarization," The Physics Classroom, n.d., https://www.physicsclassroom.com/class/light/Lesson-1/Polarization.

3. Manchester, interview, Feb. 12, 2019. The papers are V. Radhakrishnan and N. Manchester, "Detection of a Change of State in the Pulsar PSR 0833 – 45," *Nature* 222 (1969): 228–229, https://www.nature.com/articles/222228a0; and P. E. Reichley and G. S. Downs, "Observed Decrease in the Periods of Pulsar PSR 0833 – 45," *Nature* 222 (1969): 229–230, https://www.nature.com/articles/222229a0.
4. R. N. Manchester, "Pulsars at Parkes," CSIRO Astronomy and Space Science, Australia Telescope National Facility, Nov. 4, 2011, https://www.atnf.csiro.au/research/conferences/Parkes50th/ProcPapers/manchester.pdf.
5. G. Baym, in-person and email interviews with author, May 23, 2019, London.
6. J. R. Oppenheimer and G. M. Volkoff, "On Massive Neutron Cores," *Physical Review* 55 (Feb. 15, 1939).
7. V. Gribov et al., "Arkady Migdal," *Physics Today* 44, no. 12 (1991): 92.
8. "Superfluids," University of Oregon, n.d., http://abyss.uoregon.edu/~js/glossary/superfluid.html.
9. Sebastien Balibar, "The Discovery of Superfluidity," *Journal of Low Temperature Physics* 146, nos. 5–6 (Mar. 2007): 441–470.
10. Baym, interviews, May 23, 2019.
11. E. Gibney, "Neutron Stars Set to Open Their Heavy Hearts," *Nature News*, May 31, 2017, https://www.nature.com/news/neutron-stars-set-to-open-their-heavy-hearts-1.22070.
12. Baym, interviews, May 23, 2019.
13. R. R. Silbar and S. Reddy, "Neutron Stars for Undergraduates," arXiv pre-print service, Nov. 26, 2003, https://arxiv.org/pdf/nucl-th/0309041.pdf.
14. G. Ashton et al., "Rotational Evolution of the Vela Pulsar during the 2016 Glitch," *Nature Astronomy* 3 (2019): 1143–1148.
15. M. Riordan, "The Discovery of Quarks," Stanford Linear Accelerator Center, Stanford University, Apr. 1992, https://www.slac.stanford.edu/cgi-wrap/getdoc/slac-pub-5724.pdf.
16. S. Bogdanov, phone interview with author, Mar. 21, 2019.
17. J. M. Lattimer, "Neutron Star Structure and the Equation of State," *Astrophysical Journal* 550, no. 1 (2001).
18. H. T. Cromartie, "Relativistic Shapiro Delay Measurements of an Extremely Massive Millisecond Pulsar," *Nature Astronomy* 4 (2019): 72–76, https://www.nature.com/articles/s41550-019-0880-2.
19. Baym, interviews, May 23, 2019.
20. Ibid.
21. Ibid.
22. "Our History: A Passion for Discovery, a History of Scientific Achievement," Brookhaven National Laboratory, n.d., https://www.bnl.gov/about/history.

23. Baym, interviews, May 23, 2019.
24. Ibid.
25. Ibid.
26. "FAIR—The Universe in the Lab," Facility for Antiproton and Ion Research in Europe, n.d., https://fair-center.eu.
27. " Shapiro Delay," COSMOS—The SAO Encyclopedia of Astronomy, n.d., http://astronomy.swin.edu.au/cosmos/S/Shapiro+Delay.
28. T. Damour, "1974: The Discovery of the First Binary Pulsar," arXiv pre-print service, Feb. 17, 2015, https://arxiv.org/pdf/1411.3930.pdf.
29. F. Camilo, in-person interview with author, April 12, 2019, Cape Town.
30. M. Kramer, "Pulsars and General Relativity," Max Planck Institute for Radio Astronomy, Sept. 15, 2010, https://www.mpifr-bonn.mpg.de/1038767/Kramer_pulsars.pdf.
31. J. Hessels, in-person interview with author, May 10, 2019.
32. Ibid.
33. A. Watts, in-person interview with author, May 10, 2019, Amsterdam.
34. Ibid.
35. M. C. Miller et al., "PSR J0030 + 0451 Mass and Radius from *NICER* Data and Implications for the Properties of Neutron Star Matter," and T. E. Riley et al., "A *NICER* View of PSR J0030 + 0451: Millisecond Pulsar Parameter Estimation," both in *Astrophysical Journal Letters* 887, no. 1 (Dec. 12, 2019).
36. "The Neutron Star Interior Composition Explorer Mission," Goddard Space Flight Center, NASA, n.d., https://heasarc.gsfc.nasa.gov/docs/nicer.
37. Watts, interview, May 10, 2019.
38. Ibid.
39. Ibid.
40. Ibid.
41. A. Watts, "Constraining the Neutron Star Equation of State Using Pulse Profile Modelling," *AIP Conference Proceedings* 2127, no. 1 (2019).
42. K. Nandra et al., "Athena: The Advanced Telescope for High-Energy Astrophysics," European Space Agency, 2013, https://www.cosmos.esa.int/documents/400752/400864/Athena+Mission+Proposal/18b4a058-5d43-4065-b135-7fe651307c46.
43. J. Read, phone interview with author, May 6, 2019.
44. Ibid.
45. "Using Gravitational Wave Observations to Learn about Ultra-Dense Matter," LIGO, Aug. 12, 2019, https://www.ligo.org/science/Publication-GW170817ModelSelection/index.php.
46. Read, interview, May 6, 2019.

47. S. Nissanke, phone interview with author, Oct. 19, 2019.
48. K. Chatziioannou, phone interview with author, May 6, 2019.
49. Read, interview, May 6, 2019.
50. Chatziioannou, interview, May 6, 2019.
51. Ibid.

6 · How Neutron Stars Keep Spoiling Dark Matter Theories

1. "Making (Galactic) History with Big Data: First Global Age Map of the Milky Way," news release, Max Planck Institute for Astronomy, Jan. 8, 2016, https://www.mpia.de/news/science/2016-01-milky-way-agemap.
2. "South Africa's MeerKAT Telescope Discovers Giant Radio 'Bubbles' at Centre of Milky Way," news release, SKA Telescope, Sept. 12, 2019, https://www.skatelescope.org/news/meerkat-discovers-giant-radio-bubbles.
3. "Dark Energy, Dark Matter," NASA, Dec. 8, 2019, https://science.nasa.gov/astrophysics/focus-areas/what-is-dark-energy.
4. G. Bertone and D. Hooper, "A History of Dark Matter," ArXiv pre-print service, May 24, 2016, https://arxiv.org/pdf/1605.04909.pdf.
5. D. Hooper, phone and email interviews with author, Sept. 28, 2017.
6. T. Linden and B. J. Buckman, "Pulsar TeV Halos Explain the Diffuse TeV Excess Observed by Milagro," *Physical Review Letters* 120, no. 121101 (Mar. 23, 2018).
7. "Fermi Bubbles," Goddard Space Flight Center, NASA, https://fermi.gsfc.nasa.gov/science/constellations/pages/bubbles.html.
8. Hooper, interviews, Sept. 28, 2017.
9. Ibid.
10. L. Goodenough and D. Hooper, "Possible Evidence for Dark Matter Annihilation in the Inner Milky Way from the Fermi Gamma Ray Space Telescope," ArXiv pre-print service, Nov. 11, 2009, https://arxiv.org/pdf/0910.2998.pdf.
11. Hooper, interviews, Sept. 28, 2017.
12. Ibid.
13. Ibid.
14. T. Linden, phone and email interviews with author, Sept. 25, 2017, and Aug. 7, 2018.
15. Ibid.
16. Hooper, interviews, Sept. 28, 2017.
17. T. Slatyer, phone and email interviews with author, Aug. 6, 2019.
18. T. Linden, interviews, Sept. 25, 2017, and Aug. 7, 2018.
19. Ibid.
20. Slatyer, interviews, Aug. 6, 2019.

21. T. R. Slatyer et al., "The Characterization of the Gamma-Ray Signal from the Central Milky Way: A Case for Annihilating Dark Matter," *Physics of the Dark Universe* 12 (June 2016): 1–23.
22. Linden, interviews, Sept. 25, 2017, and Aug. 7, 2018.
23. Slatyer, interviews, Aug. 6, 2019.
24. C. Moskowitz, "Dark Matter May Be Destroying Itself in Milky Way's Core," *Nature News,* Apr. 8, 2014, https://www.nature.com/news/dark-matter-may-be-destroying-itself-in-milky-way-s-core-1.15018.
25. Linden, interviews, Sept. 25, 2017, and Aug. 7, 2018.
26. Slatyer, interviews, Aug. 6, 2019.
27. Ibid.
28. "National Radio Quiet Zone," National Radio Astronomy Observatory (NRAO), n.d., https://science.nrao.edu/facilities/gbt/interference-protection/nrqz.
29. "Green Bank Observatory," GBO website, n.d., https://greenbankobservatory.org.
30. Slatyer, interviews, Aug. 6, 2019.
31. R. Lynch, "The Hunt for New Pulsars with the Green Bank Telescope," arXiv pre-print service, Mar. 21, 2013, https://arxiv.org/pdf/1303.5316.pdf.
32. "Pulsar Dispersion Measure," COSMOS—The SAO Encyclopedia of Astronomy, n.d., https://astronomy.swin.edu.au/cms/astro/cosmos/p/Pulsar+Dispersion+Measure.
33. R. Lynch, in-person interview with author, Aug. 21, 2019, Green Bank Observatory, West Virginia.
34. C. Weniger et al., "Strong Support for the Millisecond Pulsar Origin of the Galactic Center GeV Excess," *Physical Review Letters* 116, no. 051102 (Feb. 4, 2016); Slatyer is among the authors of S. K. Lee et al., "Evidence for Unresolved Gamma-Ray Point Sources in the Inner Galaxy," *Physical Review Letters* 116, no. 051103 (Feb. 4, 2016).
35. Linden, interviews, Sept. 25, 2017 and Aug. 7, 2018.
36. Hooper, interviews, Sept. 28, 2017.
37. Ibid.
38. R. K. Leane and T. R. Slatyer, "Dark Matter Strikes Back at the Galactic Center," arXiv pre-print service, Apr. 19, 2019, https://arxiv.org/pdf/1904.08430.pdf.
39. Hooper, interviews, Sept. 28, 2017.
40. Ibid.
41. Ibid.
42. Slatyer, interviews, Aug. 6, 2019.
43. Ibid.

7 · When Pulsars Have Planets

1. E. Tasker, *The Planet Factory: Exoplanets and the Search for a Second Earth* (New York: Bloomsbury, 2017).
2. A. Lyne, in-person interview with author, July 12, 2019, Jodrell Bank Observatory, UK.
3. Ibid.
4. M. Bailes, in-person interview with author, Feb. 8, 2019, Parkes Telescope, Parkes, Australia.
5. Ibid.
6. Lyne, interview, July 12, 2019.
7. A. Wolszczan, phone interview with author, Aug. 30, 2019.
8. "Telescope Description," Arecibo Observatory website, n.d., https://www.naic.edu/ao/telescope-description.
9. Wolszczan, interview, Aug. 30, 2019.
10. C. DuBois, "Planets from the Very Start," Penn State University, Sept. 1, 1997, https://news.psu.edu/story/140842/1997/09/01/research/planets-very-start.
11. Wolszczan, interview, Aug. 30, 2019.
12. Ibid.
13. Ibid.
14. All quotations and observations from Wolszczan in the remainder of the chapter are from ibid.
15. A. Wolszczan and D. A. Frail, "A Planetary System around the Millisecond Pulsar PSR1257 + 12," *Nature* 355, no. 6356 (1992): 145–147.

8 · Giant Scientific Tools of the Universe

1. W. Becker, "Pulsar Timing and Its Application for Navigation and Gravitational Wave Detection," *Space Science Reviews* 214 (Feb. 2018): 30.
2. C. M. F. Mingarelli, "Probing Supermassive Black Hole Binaries with Pulsar Timing," *Nature Astronomy* 3 (2019): 8–10.
3. "South Pole Telescope," University of Chicago, Dec. 9, 2019, https://pole.uchicago.edu/spt.
4. D. Marrone, phone interview with author, Sept. 10, 2019.
5. Ibid.
6. "Astronomers Capture First Image of a Black Hole," news release, European Southern Observatory, Apr. 10, 2019, https://www.eso.org/public/unitedkingdom/news/eso1907.
7. Marrone, interview, Sept. 10, 2019.

8. "Event Horizon Telescope," Event Horizon Telescope website, n.d., https://eventhorizontelescope.org.
9. Marrone, interview, Sept. 10, 2019.
10. A. M. Ghez et al., "The Accelerations of Stars Orbiting the Milky Way's Central Black Hole," *Nature* 407, no. 6802 (Sept. 2000): 349–351.
11. "Supermassive Black Hole Sagittarius A*," NASA, Aug. 29, 2013, https://www.nasa.gov/mission_pages/chandra/multimedia/black-hole-SagittariusA.html.
12. Marrone, interview, Sept. 10, 2019.
13. Ibid.
14. "Event Horizon Telescope."
15. "Supermassive Black Hole," COSMOS—The SAO Encyclopedia of Astronomy, n.d., http://astronomy.swin.edu.au/cosmos/S/Supermassive+Black+Hole.
16. R. Pfeifle et al., "A Triple AGN in a Mid-Infrared Selected Late Stage Galaxy Merger," arXiv pre-print service, Aug. 7, 2019, https://iopscience.iop.org/article/10.3847/1538-4357/ab3a9b.
17. S. Ransom, phone interview with author, Sept. 10, 2019.
18. "Princeton Scientists Spot Two Supermassive Black Holes on Collision Course with Each Other," news release, Princeton University, July 10, 2019, https://www.princeton.edu/news/2019/07/10/princeton-scientists-spot-two-supermassive-black-holes-collision-course-each-other.
19. Ransom, interview, Sept. 10, 2019.
20. Ibid.
21. Ibid.
22. Ibid.; G. Hobbs, "Gravitational Wave Research Using Pulsar Timing Arrays," *National Science Review* 4, no. 5 (Dec. 19, 2017): 707–717.
23. H. T. Cromartie, "Relativistic Shapiro Delay Measurements of an Extremely Massive Millisecond Pulsar," *Nature Astronomy* 4 (2019), https://www.nature.com/articles/s41550-019-0880-2.
24. Ransom, interview, Sept. 10, 2019.
25. Ibid.
26. "Gravitational Wave Mission Selected, Planet-Hunting Mission Moves Forward," European Space Agency, June 20, 2017, https://sci.esa.int/web/cosmic-vision/-/59243-gravitational-wave-mission-selected-planet-hunting-mission-moves-forward.
27. M. Bailes, "MeerTime—the MeerKAT Key Science Program on Pulsar Timing," arXiv pre-print service, Mar. 18, 2018, https://arxiv.org/abs/1803.07424.
28. G. Hobbs, in-person interview with author, Feb. 12, 2019, CSIRO, Sydney.
29. M. Zumalacárregui, email interview with author, Apr. 26, 2018.

30. D. Goldberg, "Why Can't Einstein and Quantum Mechanics Get Along?" Gizmodo, Sept. 8, 2013, https://io9.gizmodo.com/why-cant-einstein-and-quantum-mechanics-get-along-799561829.
31. Zumalacárregui, interview, Apr. 26, 2018.
32. Ibid.
33. E. Siegel, "Dark Matter Winners and Losers in the Aftermath of LIGO," *Medium*, Dec. 19, 2017, https://medium.com/starts-with-a-bang/dark-matter-winners-and-losers-in-the-aftermath-of-ligo-f34ffab04fcb.
34. D. Perrodin, "Radio Pulsars: Testing Gravity and Detecting Gravitational Waves," *Physics and Astrophysics of Neutron Stars* (Jan. 10, 2019): 95–148.
35. A. Archibald, email interviews with author, Aug. 22, 2017, and Sept. 15, 2019.
36. M. Kramer, phone and email interviews with author, Aug. 31, 2017.
37. A. Lyne, in-person interview with author, July 12, 2019, Jodrell Bank Observatory, UK; M. Burgay, "The Double Pulsar System in Its 8th Anniversary," arXiv pre-print service, Oct. 3, 2012, https://arxiv.org/abs/1210.0985.
38. Lyne, interview, July 12, 2019.
39. Ibid.
40. Burgay, "Double Pulsar System."
41. Archibald, interview, Aug. 22, 2017.
42. Kramer, interviews, Aug. 31, 2017.
43. S. Ransom et al, "A Millisecond Pulsar in a Stellar Triple System," *Nature* 505 (Jan. 23, 2014): 520–524.
44. R. Lynch, in-person interview with author, Aug. 21, 2019, Green Bank Observatory, Green Bank, West Virginia.
45. Ibid.
46. Archibald, interviews, Aug. 22, 2017, and Sept. 15, 2019.
47. Ibid.
48. Lynch, interview, Aug. 21, 2019.
49. "Equivalence Principle," Encyclopaedia Britannica website, n.d., https://www.britannica.com/science/equivalence-principle.
50. Archibald, interviews, Aug. 22, 2017, and Sept. 15, 2019.
51. Ibid.
52. Ibid.
53. Lynch, interview, Aug. 21, 2019.
54. Archibald, interviews, Aug. 22, 2017 and Sept. 15, 2019.
55. Zumalacárregui, interview, Apr. 26, 2018.
56. Archibald, interviews, Aug. 22, 2017, and Sept. 15, 2019.
57. Ibid.
58. Ibid.

9 · Fast Radio Bursts, an Unfinished Chapter

1. D. Lorimer, phone and email interviews with author, Apr. 10, 2019.
2. Ibid.
3. D. Cossins, "Fast Radio Bursts: We're Finally Decoding Messages from Deep Space," *New Scientist,* May 8, 2019, https://www.newscientist.com/article/mg24232291-900-fast-radio-bursts-were-finally-decoding-messages-from-deep-space.
4. Lorimer, interviews, Apr. 10, 2019.
5. Ibid.
6. E. Keane, "High Time-Resolution Astrophysics," Jodrell Bank Observatory, University of Manchester, Apr. 17, 2008, http://www.jb.man.ac.uk/~ekean/my_damtp_presentation.pdf.
7. Lorimer, interviews, Apr. 10, 2019.
8. Ibid.
9. J. O'Callaghan, "Mysterious Outburst's Quiet Cosmic Home Yields More Questions Than Answers," *Scientific American,* June 27, 2019, https://www.scientificamerican.com/article/mysterious-outbursts-quiet-cosmic-home-yields-more-questions-than-answers.
10. Lorimer, interviews, Apr. 10, 2019.
11. Ibid.
12. M. Bailes, interview with author, Feb. 10, 2019.
13. A. Loeb, phone and email interviews with author, Apr. 11, 2019.
14. K. Kellermann and B. Sheets, *Serendipitous Discoveries in Radio Astronomy* (Green Bank, WV: National Radio Astronomy Observatory, May 4–6, 1983).
15. Loeb, interviews, Apr. 11, 2019.
16. C. Woolston, "Microwave Oven Blamed for Radio-Telescope Signals," *Nature,* May 8, 2015, https://www.nature.com/news/microwave-oven-blamed-for-radio-telescope-signals-1.17510.
17. H. W. Lin, G. Gonzalez Abad, and A. Loeb, "Detecting Industrial Pollution in the Atmospheres of Earth-Like Exoplanets," *Astrophysical Journal Letters* 792, no. 1 (Aug. 12, 2014).
18. D. Thornton et al., "A Population of Fast Radio Bursts at Cosmological Distances," *Science* 340, no. 6141 (July 5, 2013), https://arxiv.org/ftp/arxiv/papers/1307/1307.1628.pdf.
19. Cossins, "Fast Radio Bursts."
20. Loeb, interviews, Apr. 11, 2019.
21. "Breakthrough Initiatives," Breakthrough Initiatives website, n.d., https://breakthroughinitiatives.org/.
22. S. Chatterjee, email interview with author, Apr. 6, 2017.

23. S. Chatterjee, "Focus on the Repeating Fast Radio Burst FRB 121102," *Astrophysical Journal*, n.d., https://iopscience.iop.org/journal/0004-637X/page/Focus_on_FRB_121102.
24. Chatterjee, interview, Apr. 6, 2017.
25. Ibid.
26. Loeb, interviews, Apr. 11, 2019.
26. "Breakthrough Initiatives."
27. Chatterjee interview, Apr. 6, 2017.
28. "The Chemical Composition of the Universe," *COSMOS—The SAO Encyclopedia of Astronomy*, n.d., http://astronomy.swin.edu.au/cosmos/C/Chemical+Composition.
29. Chatterjee, interview, Apr. 6, 2017.
30. Ibid.
31. D. Lorimer, "Fast Radio Bursts: Nature's Latest Cosmic Mystery," Aspen Center for Physics, n.d., http://aspen17.phys.wvu.edu/Lorimer.pdf.
32. "Could Fast Radio Bursts Be Powering Alien Probes?" news release, Harvard and Smithsonian Center for Astrophysics, Mar. 9, 2017, https://www.cfa.harvard.edu/news/2017-09.
33. Chatterjee, interview, Apr. 6, 2017.
34. "Breakthrough Listen Detects Repeating Fast Radio Bursts from the Distant Universe," news release, Breakthrough Initiatives, Aug. 29, 2017, https://breakthroughinitiatives.org/news/13.
35. "Breakthrough Listen, the World's Biggest SETI Program, to Incorporate the Southern Hemisphere's Biggest Radio Telescope—the MeerKAT array—in Its Existing Search for Extraterrestrial Signals & Technosignatures," news release, Breakthrough Initiatives, Oct. 2, 2018, https://breakthroughinitiatives.org/news/23.
36. "The Canadian Hydrogen Intensity Mapping Experiment Is a Revolutionary New Canadian Radio Telescope Designed to Answer Major Questions in Astrophysics & Cosmology," CHIME website, n.d., https://chime-experiment.ca.
37. V. Kaspi, email interview with author, Mar. 15, 2020.
38. Ibid.
39. Ibid.
40. "Over $100M in Research Infrastructure Support to McGill," press release, McGill University, May 29, 2015, https://www.mcgill.ca/newsroom/channels/news/over-100m-research-infrastructure-support-mcgill-253109.
41. Kaspi, interview, Sept. 18, 2019.
42. Ibid.
43. R. Shannon, in-person interview with author, Feb. 5, 2018, Swinburne University, Melbourne, Australia.

44. Ibid.
45. "The Australian Square Kilometre Array Pathfinder (ASKAP) Telescope," CSIRO, https://www.csiro.au/en/Research/Facilities/ATNF/ASKAP.
46. Shannon, interview, Feb. 5, 2018.
47. A. Deller, email interview with author, Mar. 20, 2020.
48. Shannon, interview, Feb. 5, 2018.
49. Ibid.
50. R. Shannon, *A Fly's Eye FRB Survey with ASKAP*, Swinburne University and Ozgrav, June 2018, http://caastro.org/wp-content/uploads/2018/06/Shannon-FRB2018.pdf.
51. Shannon, interview, Feb. 5, 2018.
52. T. Stephens, "Astronomers Make History in a Split Second with Localization of Fast Radio Burst," news release, University of California Santa Cruz, June 27, 2019, https://news.ucsc.edu/2019/06/fast-radio-burst.html.
53. Shannon, interview, Feb. 5, 2018.
54. Deller, interview, Mar. 20, 2020.
55. Shannon, interview, Feb. 5, 2018.
56. W. Clavin, "Fast Radio Burst Pinpointed to Distant Galaxy," news release, California Institute of Technology, July 2, 2019, https://www.caltech.edu/about/news/fast-radio-burst-pinpointed-distant-galaxy.
57. Shannon, interview, Feb. 5, 2018.
58. M. Amiri et al., "A Second Source of Repeating Fast Radio Bursts," *Nature* 566 (2019): 235–238, https://www.nature.com/articles/s41586-018-0864-x.
59. B. C. Andersen et al., "CHIME/FRB Detection of Eight New Repeating Fast Radio Burst Sources," *Astrophysical Journal Letters* 885, no. 1 (October 31, 2019), https://iopscience.iop.org/article/10.3847/2041-8213/ab4a80; E. Fonseca et al., "Nine New Repeating Fast Radio Burst Sources from CHIME/FRB," *Astrophysical Journal Letters* 891, no. 1 (Feb. 26, 2020), https://iopscience.iop.org/article/10.3847/2041-8213/ab7208.
60. Lorimer, interviews, Apr. 10, 2019.

Acknowledgments

This book is dedicated to my two sons, Tima and Kai. I really hope that it will spark in you a fascination for our enigmatic world—and a drive to disentangle the mesmerizing web of secrets that the Universe holds.

It's also dedicated to Fritz Zwicky. Without his genius, we might never have discovered the enthralling link between the teeny tiny neutron, the stunning clouds in deep space—supernovas—and the most mind-blowing, ultra-dense spinning spheres that pop into existence when a massive star goes out with a boom: neutron stars.

I'd like to say a huge thank you to many, many scientists who met with me on my travels, and to those who found the time to fact-check the manuscript before publication. In particular, I'd like to thank researchers Kate Alexander, Anne Archibald, Will Armentrout, Matthew Bailes, Gordon Baym, Jocelyn Bell Burnell, Edo Berger, Slavko Bogdanov, Marica Branchesi, Rene Breton, David Buckley, Fernando Camilo, Shami Chatterjee, Katerina Chatziioannou, Jim Cordes, Adam Deller, Ron Ekers, Jason Hessels, Anna Ho, George Hobbs, Dan Hooper, Thomas Janka, Vicky Kaspi, Stavros Katsanevas, Chryssa Kouveliotou, Michael

Kramer, Tom Landecker, Tim Linden, Avi Loeb, Malcolm Longair, Dunc Lorimer, Ryan Lynch, Andrew Lyne, Dick Manchester, Dan Marrone, Chia Min Tan, Samaya Nissanke, Scott Ransom, Jocelyn Read, Anish Roshi, John Sarkissian, Ryan Shannon, Tracy Slatyer, Stephen Smartt, Chris Thompson, Alan Vaughan, Anna Watts, Alex Wolszczan, Ofer Yaron, Miguel Zumalacárregui—and many more. I'd also like to thank all the outreach people, including Angus Flowers, who drove me to the SKA site in South Africa and organized helpful interviews, Frank Nuijens from ASTRON in the Netherlands, who took me to the swampy field where LOFAR is located, Rob Hollow from CSIRO, who helped me with interviews in Australia, Natalie Butterfield from NRAO, who assisted me with the Green Bank Telescope visit, and many others.

I'd also like to thank my husband, Tim Weber, and my former WIRED UK colleague Matt Reynolds for helping me to edit the early drafts; former Harvard University Press editor Jeff Dean, who asked whether I wanted to write such a book in the first place; my editor James Brandt and copy editor Julie Carlson for all their great, insightful advice during the editing process; and all the others who have helped me with various aspects of the book and all the traveling.

During my travels, I was thrilled to see the sheer numbers of talented female astronomers, astrophysicists, and gravitation wave researchers. Astronomy and physics used to be a profession dominated by men. We are still far from parity—but there are more and more amazing women exploring space, and being a female physicist myself, I'm honored to count them among my heroes.

There is, of course, the remarkable dark matter detective Vera Rubin, and the aliens chaser Jill Tarter (in the film *Contact,* shot at Arecibo Telescope in Puerto Rico, which I refer to in this book, the main character played by Jodi Foster is based on Jill). There are comet hunters Carolyn Shoemaker and, two centuries prior, Caroline Herschel. There are Carolyn Porco, one of the world's leading expert on planetary rings, Margaret Geller, who studies the structure of galaxies and the distribution of dark matter in them, and many more female astronomers and astrophysicists. And of course, all the amazing pulsar astronomers whom I met on my travels or interviewed on the phone—Kate Alexander, Anne Archibald, Jocelyn Bell Burnell, Marica Branchesi, Katerina Chatziioannou, Thankful Cromartie, Anna Ho, Vicky Kaspi, Chryssa Kouveliotou, Samaya Nissanke, Jocelyn Read, Tracy Slatyer, Renee Spiewak, Anna Watts, and many, many more.

Then there are all the amazing, fearless female astronauts, including the Soviet cosmonaut Valentina Tereshkova, who in 1963 became the first woman to go to space; Sally Ride, the first American woman in space; Mae Jemison, the first black female astronaut to go to space, and many others. For me, one person, though, is especially important and always will be. It's the second Canadian woman to go to space, astronaut Julie Payette, who's now the twenty-ninth governor general of Canada. When I was a teenager in the late 1990s, wearing baggy jeans, watching Beverly Hills 90210, and listening to Michael Jackson, one day I saw her portrait displayed on the wall of my high school—College Mont-Saint-Louis in the north of Montreal. It turned out that she went there too, and the school is extremely proud of her, as it should be. When I found out who she was and learned that she was just about to go to space on May 27, 1999, the year I graduated from high school, I suddenly wanted to do just that—I wanted to be an astronaut. So I decided to write her a letter (a real letter—no email at the time!), asking for her advice on what to study at university. I didn't think she'd reply, but she did: she wrote me a postcard back, a Christmas postcard featuring a virtual Christmas tree added to the International Space Station. I still count the card among my most precious possessions.

She wrote: "I really appreciate your letter and your desire to become an astronaut. That's what I also wanted at your age. The most important thing is to choose a science area that you like, doesn't matter what it is, and to excel in it. And then you just have to apply to the Canadian Space Agency during the next recruitment process.

With a bit of effort and discipline, one day, the world will be at your feet."

Even though I never became an astronaut (partly because my eyesight wasn't good enough to become a jet pilot), she inspired me to go into engineering (like her) and later astrophysics. But most of all, thanks to her I fell in love with space—forever. Thank you, Julie.

Readers, I hope you love space as much as I do and that you've enjoyed traveling with me around the world and beyond, far, far away into the Universe. And I hope that this book will inspire younger readers to dive into whatever area of science they might choose. Because we need people who get it, and people who dare.

And especially people who, from time to time, like to look up.

Index

accretion, 11, 100–104, 129, 197
accretion disks, 11, 197
accretion-powered pulsars, 99–105
active galactic nuclei (AGNs), 40, 41, 172, 211, 242
Advanced LIGO (Laser Interferometer Gravitational Wave Observatory), 18–22
Advanced Virgo, 18–22
advance of periastron, 229–230
AGNs (active galactic nuclei), 40, 41, 172, 211, 242
Alexander, Kate, 27–28, 30
alien civilization signaling, theories of, 47, 235
ALMA (Atacama Large Millimeter/submillimeter Array), 62–64, 82–84, 93
Alpha Centauri, proposed transportation of humans to, 237, 240
alternative gravity theories, 214–217
Anderson, Phil, 137
Andromeda, 164, 165, 205
Anglo Australian Telescope, Vela Pulsar observed with, 75

Anomalous X-ray Pulsars (AXPs), 117–121
antimatter, 168–169
Archibald, Anne, 104, 218, 222–226
Arecibo Telescope: Cordes on, 57; Crab Nebula observed with, 75–76; design and construction of, 188–192; limitations of, 244; message sent into space by, 190–191; millisecond pulsars observed with, 100, 104, 118; multibeam receiver at, 128; PALFA (Pulsar Arecibo L-band Feed Array), 238–242; pulsing timing with, 130, 209; scientific achievements associated with, 192–197; testing of general relativity with, 228
arXiv.org, 173, 176
ASCA satellite, 114
ASKAP (Australian Square Kilometer Array Pathfinder), 151, 244, 250–254
Asteroid Terrestrial Impact Last Alert System (ATLAS), 63
ASTRON (Netherlands Institute for Radio Astronomy), 92–93. *See also* LOFAR (Low Frequency Array)
AT2018cow, 63–64, 82–84

Atacama Desert, Chile: ALMA (Atacama Large Millimeter/submillimeter Array), 62–64, 82–84, 93; landscape and terrain, 60–61; Las Campanas Observatory in, 26, 77; Paranal Observatory in, 61
ATLAS (Asteroid Terrestrial Impact Last Alert System), 63
atomic nucleus, 134–135
atomic theory, development of, 35–37, 67–68
Australian Square Kilometer Array Pathfinder (ASKAP), 151, 244, 250–254
AXPs (Anomalous X-ray Pulsars), 117–121

Baade, Walter, 38, 68–69
Backer, Donald C., 20, 99, 104, 154, 210
Bahcall, John, 188, 196
Bailes, Matthew, 1–3; Breakthrough Listen project, 244; Loeb and, 236, 240; Lorimer and, 234–235; Molonglo Observatory Synthesis Telescope and, 74; paper retracted by, 186–188; Parkes Observatory and, 52–56, 128, 249
Baksan neutrino detector, 77
Barish, Barry C., 6, 19
baryons, 141, 164
Bassa, Cees, 99, 104
Baym, Gordon, 133–139, 144–148
beacons, pulsars as, 198–200
Bell, Jocelyn: career and awards, 49; Hewish's supervision of, 44–46, 48; Interplanetary Scintillation Array constructed by, 41–43; LGM-1 pulsar discovered by, 1–2, 36, 44–49, 72–73, 95, 235; Manchester on, 132; marriage, 48–49; quasar research by, 43–44
Bell Labs, 36
Berger, Edo, 26–29
Betelgeuse, 80
Bethe, Hans, 71, 144–145
beyond-Horndeski theories, 216–217
Big Bang: antimatter produced by, 168–169; cosmic microwave background radiation from, 36, 200; elements formed after, 31–33, 241; matter and antimatter produced from, 168–169; quark-gluon plasma resulting from, 145
binary black hole mergers, 19
binary neutron stars, 4–8, 11, 23–31
binary pulsars, 192
binary triple system, 222–228

binding energy, pulsar, 226
Bjorken, James, 145
blackbodies, 33
black dwarfs, 65
BlackHoleCam, 228
black holes: Chandrasekhar's theory of, 66–67; detection and observation of, 204–205; event horizon, 202–205; formation of, 11; mergers of, 6, 19, 22–23, 199, 205–207; Schwarzschild radius of, 204; stellar-mass, 200–207; supermassive, 203–205. *See also* Sagittarius A*
"black widow" pulsars, 104
blazars, 7–8
Blewett, John, 146
blue supergiants, 77–82, 87–89
Bohr, Niels, 70
Bolton, John, 39–40
Boschi, Valerio, 15, 16
Brahe, Tycho, 70
Branchesi, Marica: GW170817 merger detected by, 5–8, 25–31, 206; marriage and domestic life, 5, 6–7; multi-messenger astronomy project, 19–22; *Time* magazine recognition of, 8
Brans-Dicke gravity, 216
Breakthrough Initiatives, 237–238, 243
Breakthrough Listen project, 237–238, 243
Breakthrough Starshot project, 237
Breton, Rene, 221
Bruno Rossi Prize, 172
Buckley, David, 102–103
Burgay, Marta, 220
Burrows, Adam, 110

Cagliari Observatory, 220
California Institute of Technology (Caltech): Deep Synoptic Array 10, 244; Fuller at, 81; Owens Valley Radio Observatory at, 253; Zwicky at, 37–38, 67, 164–165
Cambridge University: acceptance of women at, 48–49; Cambridge Observatory at, 130; Cavendish Laboratory at, 35–37, 135
Cameron, Alastair, 134
Camilo, Fernando, 121–125, 153
Canada Foundation for Innovation, 246
Canadian Hydrogen Intensity Mapping Experiment (CHIME), 245–250
Cantiello, Matteo, 76

Cassini space probe, 218
Cassiopeia, 48, 70, 117
Cassiopeia A, 85
Catalogue of Galaxies and of Clusters of Galaxies (Zwicky), 73
Cavendish Laboratory, 35–37, 135
CBM (compressed baryonic matter) experiment, 147–148
Centre National de la Recherche Scientifique (CNRS), 15
CERN, Large Hadron Collider, 146–147, 166
Chads, 44
Chadwick, James, 36, 37–38, 66, 67–68
Chakrabarty, Deepto, 118–119
Chandrasekhar, Subrahmanyan, 64–67
Chandra X-ray Observatory, 29, 101, 102, 124–125
Chatterjee, Shami, 238–245
Chatziioannou, Katerina, 158–159, 162
CHIME (Canadian Hydrogen Intensity Mapping Experiment), 244–250
chirp mass, 160
circumstellar material, 79–80
Clifton, Trevor, 186
CNRS (Centre National de la Recherche Scientifique), 15
cobalt, origin of, 32–33
Collins, Robin, 46
Coma cluster, 165, 214
Commensal Real-time ASKAP Fast Transients (CRAFT), 250
compressed baryonic matter (CBM) experiment, 147–148
Contact (film), Arecibo Telescope featured in, 192
cooling of neutron stars, 137
Cordes, Jim, 56, 58, 239
core-collapse supernovas, 77–82, 87–89
core structure, neutron star: inner, 141–148; outer, 139–141
Corsi, Alessandra, 30
Cosmic Explorer, 18
cosmic microwave background radiation, 36, 200
Coulter, Dave, 26
"Cow" supernova, 63–64, 82–84
Crab Nebula, 51, 70, 75–76, 192
Crab Pulsar, 51, 85, 98, 233

CRAFT (Commensal Real-time ASKAP Fast Transients), 250
Cromartie, Thankful, 210
crust, neutron star, structure of, 137–138
CSIRO, 55, 72, 213
Cygnus X-1, 102, 205

Dark Energy Camera, 28
dark matter: dark matter haloes, 166; evidence for, 163–166; in Galactic Center, 163–164; gamma-ray excess and, 171–177; gravitational lensing and, 166; Hooper's theory of, 171–177, 182–183; positron excess and, 167–171; prevalence of, 164; Rubin and Ford's theory of, 165; Zwicky's theory of, 72, 164–165, 214
dark matter haloes, 166
"death line," 91
Deep Space Network, 55, 199
Deep Synoptic Array, 253
Deep Synoptic Array-10, 244, 253
Deller, Adam, 250–253
Delta parameter, 227
density: of neutron stars, 9, 11, 71–72, 138–139; of white dwarfs, 65–67
Detweiler, Steven, 207–208
discovery of neutron stars: Bell's observations, 1–2, 36, 44–49, 72–73, 95, 235; Bolton and Hazard's observations, 39–40; impact of, 255–257; Interplanetary Scintillation Array and, 41–49; Parkes Observatory and, 51–56; Schmidt's observations, 40; Zwicky's theory of, 38, 50, 67–73
dispersion, 58
Distance Less Than 40 Mpc survey, 28
Dobbs, Matt, 246, 248
Dominion Radio Astrophysical Observatory (DRAO), 245. *See also* CHIME (Canadian Hydrogen Intensity Mapping Experiment)
Doppler effect, 202
Double Pulsar, 55, 105, 126, 153, 220, 222
Downs, Gabriel, 208
Drake, Frank, 179
Drever, Ronald, 6, 14
Duhalde, Oscar, 77
Duncan, Rob, 110–117
dwarf galaxies, 183, 241
Dwingeloo Radio Observatory, 92

dynamo effect, 110–111
Dyson, Frank Watson, 130

Earth: Earth–Moon–Earth communication, 92; Earth–Moon–Sun system, 225–226; magnetic field of, 111
Échelle Atomique Libre, 213
Eddington, Arthur Stanley, 65–67, 130
Eddington limit, 108, 129
Effelsberg Radio Telescope, 210
EHT (Event Horizon Telescope), 200–204, 228
Einstein, Albert, 6, 12–13, 130, 192, 229–230. *See also* general theory of relativity
Einstein Telescope, 18
Einstein X-ray Observatory, 102, 116–117
Ekers, Ron, 72–73
electromagnetic radiation, emission of, 9, 22, 39, 171, 178
electromagnetic spectrum, 39
electrons: in core-collapse, 88, 95; discovery of, 36
elements, formation of, 11, 31–33, 87–88, 241
enhanced X-ray Timing and Polarimetry (eXTP), 157
Ensemble Pulsar Scale, 213
Epsilon Eridani, observation of, 179
equation of state, 142–144, 161–162
equivalence principle, 214, 222–228
European Gravitational Observatory (EGO), 15
European Pulsar Timing Array (PTA), 210
European Space Agency, 83; INTEGRAL telescope, 83, 103; LISA (Laser Interferometer Space Antenna), 212, 228, 256
eV (electron-volts), 167
event horizon, 202–205
Event Horizon Telescope (EHT), 200–204
extragalactic magnetars, 239

Facility for Antiproton and Ion Research (FAIR), 147–148, 256
Fagg, Harry, 54
Fahlman, Greg, 117
FAIR (Facility for Antiproton and Ion Research), 147–148, 256
fallback disk, 197
Faraday cage, 149
Faraday rotation, 243

FAST (Five-hundred-meter Aperture Spherical radio Telescope), 128, 130, 212, 228
fast radio bursts (FRBs), detection of, 4; ASKAP (Australian Square Kilometer Array Pathfinder), 250–254; Breakthrough Listen project, 235–238; CHIME (Canadian Hydrogen Intensity Mapping Experiment), 244–250; PALFA collaboration, 238–245; Parkes Multibeam Pulsar Survey of the Magellanic Clouds, 231–235; with Parkes Multibeam Survey of the Galactic Plane, 126; SETI theory of, 235–236, 254
Fermi bubbles, 172, 173–177
Fermi Gamma-Ray Space Telescope: gamma-ray flash observed by, 29–30; GW170817 merger detected by, 12, 24; J0002+6126 observed by, 85; microwave haze observed by, 171–172
Fermi haze, 171–172
fermions, Pauli exclusion principle for, 65
Fernandez, Rodrigo, 34
Finkbeiner, Doug, 171–175
Finkelstein, David, 204
Five-hundred-meter Aperture Spherical radio Telescope (FAST), 128, 130, 212, 228
Flowers, Angus, 149–151
flux density, 123
Foley, Ryan, 26
Fonseca, Emmanuel, 210
Ford, Kent, 165
formation of neutron stars: debris clouds resulting from, 11; elements formed by, 11, 31–33; supernova explosions and, 3–4, 9–10; tidal effects in, 10–11
Fowler, Ralph H., 66
Frail, Dale, 195–196, 230
FRBs. *See* fast radio bursts (FRBs), detection of
Fuller, Jim, 81

Galactic Center: dark matter in, 163–164, 166–167; gamma-ray excess in, 171–177; MeerKAT's imaging of, 150; PSR J1745−2900 in, 125; pulsars in, 180–184. *See also* Sagittarius A*
Galactic Plane: gamma radiation in, 167–168; Parkes Multibeam Survey of, 126; Pulsar Arecibo L-band Feed Array (PALFA), 238–242

galaxies: active galactic nuclei, 40, 211; collision and merger of, 199, 205–206; dwarf, 183, 241; galaxy NGC 4993, 28; mergers of, 205; number of, 76, 126; radio, 41; Virgo Cluster, 15; Zwicky's catalogue of, 73, 164. *See also* Big Bang; black holes; dark matter; Milky Way

Galilei, Galileo, 14

Galileo navigation system, 198

gamma-ray bursts (GRBs), 105–110

gamma rays: emission of, 11, 20, 29–30, 96–97; excess in, 171–177, 182–184; in Galactic Plane, 167–168; gamma-ray bursts, 105–110; GCN circulars, 25; observation and detection of, 24–30; wavelength of, 33. *See also* magnetars

Gavriil, Fotis, 119

GBT. *See* Green Bank Telescope (GBT)

GBT 350 MHz Drift-Scan Survey, 222–223

GCN (gamma-ray burst coordinates) circulars, 25–31

Geminga pulsar, 170

general theory of relativity: black holes and, 200, 202; concept of, 12–14; confirmation of, 192; core collapse and, 68; dark matter and, 164; equivalence principle of, 222–228; geodetic precession in, 219; neutron star observations and, 2, 11, 31; post-Keplerian parameters and, 228–230; pulsar tests of, 130, 217–222; quantum mechanics and, 214–215; Shapiro delay predicted by, 228–229; for spherical mass, 204; triple system and, 226–228

Genzel, Reinhard, 203

geodetic precession, 219

GeV (gigaelectron-volts), 173

Giaccon, Riccardo, 101

gigaelectron-volts (GeV), 173

"glitches," 118–120, 133–141

global positioning systems (GPS), pulsars as beacons for, 198–200

globular clusters, millisecond pulsars in, 181

GLONASS, 198

gluons, 141

gold, origin of, 11, 31–33

Gold, Thomas (Tommy), 50–51

GoldenEye (film), Arecibo Telescope featured in, 190

Goodenough, Lisa, 172–174

gravitational lensing, 166

Gravitational Wave Inaf Team (GRAWITA), 26

gravitational waves: black hole mergers and, 206–207; detection of, 200–207; equivalence principle, 214; Hulse-Taylor pulsar system, 13–14; neutron star mergers and, 215–222; observation and detection of, 6, 14–18, 206, 211–212, 255; origins of, 10; speed of, 217. *See also* gravity, theories of; multi-messenger astronomy

gravity, theories of: Brans-Dicke, 217; Horndeski and beyond-Horndeski, 216; modified Newtonian dynamics, 216; neutron star mergers and, 215–222; Newton's theory of, 12, 214; TeVeS (tensor-vector-scalar), 216. *See also* general theory of relativity; gravitational waves

GRAWITA (Gravitational Wave Inaf Team), 26

GRBs (gamma-ray bursts), 105–110

Green Bank Interferometer (GBI), 179

Green Bank Telescope (GBT): Breakthrough Listen influence at, 243; Crab Nebula observed with, 75; design and construction of, 177–180; GBT 350 MHz Drift-Scan Survey, 222–223; history of, 179; limitations of, 244; PSR J0740 + 6620 detected with, 143; PSR J1023 + 0038 detected with, 104; PSR J1748 – 2446ad detected with, 154; pulsars observed with, 180–182; pulsar timing with, 130, 209; testing of general relativity with, 228

Gregory, Philip, 117

guest stars, 4

GW170817 merger, 4–12, 23–31, 158–161

Hale Telescope, 40

Halpern, Jules, 121–122

Halpern, Mark, 246

Hanna, Chad, 24

Harms, Jan, 5

HAWC (High-Altitude Water Cherenkov Observatory), 170–171

Hawking, Stephen, 237

Hazard, Cyril, 39–40

heavy ion collisions, 145–148

Helios 2 spacecraft, 107

helium: after Big Bang, 241; in core-collapse supernovas, 87; superfluid state of, 135–136
helium nuclei, formation of, 32
Hellings, Ronald, 207–208
Hellings-Downs theory, 209
Hessels, Dimphy, 99–100
Hessels, Jason, 99, 154–155
Hewish, Anthony, 41–49, 73, 132, 152. *See also* Interplanetary Scintillation Array
High-Altitude Water Cherenkov Observatory (HAWC), 170–171
high mass x-ray binaries (HMXBs), 103
Hinshaw, Gary, 246
Ho, Anna, 63–64, 81, 82–84
Hobbs, George, 213
Hooper, Dan: dark matter theory of, 171–177, 182, 184; positron excess observed by, 167–171
Horndeski, Gregory, 216
Horndeski theories, 216, 217
hot spots, 101, 155–158
Howard E. Tatel telescope, 179
Hubble Space Telescope, 29, 236
Hulse, Russell Alan, 13, 152, 192, 206, 229
Hulse-Taylor pulsar, 13–14, 152, 218–219, 222, 230
hydrogen: in constellation Ophiuchus, 180; formation of, 87, 241
hyperons, 142, 143

IceCube Observatory, 8, 22
IMB (Irvine-Michigan-Brookhaven) neutrino detector, 77
IMXBs (intermediate mass x-ray binaries), 103
Indigo-Indian gravitational wave consortium, 23
inertia, moment of, 133
INFN (Istituto Nazionale di Fisica Nucleare), 15
inner core, composition of, 141–148
INTEGRAL telescope, 83, 103
Intelligent Life in the Universe (Shklovsky), 134
intermediate mass x-ray binaries (IMXBs), 103
Internal Constitution of the Stars, The (Eddington), 65
International Atomic Time, pulsars in cross-checking of, 10
International Sun-Earth Explorer 1 (ISEE-3), 107
International Union of Radio Science (URSI), 151

Interplanetary Scintillation Array: construction and design of, 35–36, 41–43; pulsars detected with, 43–49
interstellar medium, 30, 40–41, 56–59
Ioffe Institute, 108
iron: in core-collapse supernovas, 87–88; in neutron star crust, 137–138; origin of, 32–33
Irvine-Michigan-Brookhaven (IMB) neutrino detector, 77
ISABELLE, 146
ISEE-3 (International Sun-Earth Explorer 1), 107
Istituto Nazionale di Fisica Nucleare (INFN), 15
Iyer, Bala, 23

J0002+6126, 85
Janka, Thomas, 85–86
Jansky, Karl, 203
Jeanette (Parkes Observatory cook), 53–54
Jet Propulsion Laboratory (JPL), Vela glitch observed at, 133–134
Jodrell Bank Observatory, 1, 127, 186
Jupiter, radiation belt of, 179

Kamiokande II neutrino detector, 77
kaons, 142–143
Kapitsa, Pyotr, 70, 135–136
Karl G. Jansky Very Large Array, 179
Kaspi, Vicky, 118–120, 222, 245–250
Katsanevas, Stavros, 15–16
KAT 7, 148
Keck Telescope, 63, 79
Kepler, Johannes, 70
Kepler's laws, 228–230
Kepler space telescope, 197
kilonovas: color of, 33–34; emission of, 11, 20; observation and detection of, 28–29
Kouveliotou, Chryssa, 105–110, 113–118
Kramer, Michael, 153, 218–219, 222
Kulkarni, Shri, 99

Lambda point, 135–136
Landau, Lev, 38, 70–71, 135–136
Langer, Norbert, 80
Large, Michael, 74–75
Large Hadron Collider (LHC), 146–147, 166
Large Magellanic Cloud (LMC), 55, 77, 108

Large Synoptic Survey Telescope, 89
Large Telescope Working Group, 151
Las Campanas Observatory, 26, 77
Las Cumbres Observatory, 28
Laser Interferometer Gravitational Wave Observatory. *See* LIGO (Laser Interferometer Gravitational Wave Observatory)
Laser Interferometer Space Antenna (LISA), 212, 228, 256
La Silla Observatory, 26
Lattimer, James, 110
Law, Casey, 240
Leane, Rebecca, 183
LGM-1 pulsar: Bell's discovery of, 1–2, 44–49, 72–73, 95; Manchester's observations of, 131–133
LHC (Large Hadron Collider), 146–147, 166
light cylinder, 97–98
LIGO (Laser Interferometer Gravitational Wave Observatory): Advanced LIGO upgrades to, 18–19; black hole merger detected by, 19, 22–23; construction of, 16–17; cost of, 15; detection and observation of GW170817, 10; funding and construction of, 14; GW170817 merger detected by, 2, 4–12, 23–31, 158–161; limitations of, 206–207; in multi-messenger astronomy project, 19–22
LIGO Scientific Collaboration, 14
Linden, Tim, 174–176, 182
LISA (Laser Interferometer Space Antenna), 212, 228, 256
lithium, formation of, 241
Liverpool telescope, 63
LMC (Large Magellanic Cloud), 55, 77, 108
LMXBs (low mass x-ray binaries), 101–103
Loeb, Avi, 235–243
LOFAR (Low Frequency Array), 90–95
LOFAR Tied-Array All-Sky Survey (LOTAAS), 90–91
Longair, Malcolm: on Chadwick, 37; on discovery of pulsars, 47, 49–50; on Interplanetary Scintillation Array, 43; leadership of Cavendish Laboratory, 35, 37; on Oppenheimer, 38–39; quasar research of, 40; on Rutherford, 37; on Zwicky, 38
Lorimer, Duncan, 220, 231–235, 250

Lorimer burst, 126
Los Alamos National Laboratory, magnetars detected by, 106
LOTAAS (LOFAR Tied-Array All-Sky Survey), 90–91
Lovelace, Richard, 75–76
Lovell Telescope, 1; magnetars observed with, 125; PSR B1829 – 10 detected by, 186; pulsar detection and observation with, 55, 104; pulsar timing with, 130, 210
Low Frequency Array (LOFAR), 90–95
low frequency radio waves, 94–95
low mass x-ray binaries (LMXBs), 101–103
lunar occultation technique, 39–40
Luyten, Willem, 65
Lynch, Ryan, 180–181, 210, 212, 222–228
Lyne, Andrew, 121, 127–128, 185–188, 196, 220–221, 227

M13, Arecibo message broadcast to, 190–191
magnesium, formation of, 87
magnetars, 105–118; amplification of magnetic field in, 111–113; Anomalous X-ray Pulsars, 117–121; definition of, 4; dynamo theory of, 110–111; extragalactic, 239; fading and revival of, 124–125; fast radio bursts and, 241–242; magnetic reconnection in, 113; PSR J1745 – 2900, 125; radio wave emissions by, 121–124; soft gamma repeaters and, 105–110, 114–116; spin-down rate of, 113–119; starquakes and, 112; timing of, 118–120, 245; x-ray emissions by, 105–106, 123–124; XTE J1810 – 197, 121–125
magnetic fields: dynamo effect, 110–111; of Earth, 111; Faraday rotation and, 243; of interstellar medium, 57; of neutron stars, 57, 97–98, 169; of Sun, 111. *See also* magnetars
magnetic flux, 97
magnetic reconnection, 113
magnetosphere, 51
"Magnificent Seven," 96–97
Manchester, Dick, 55, 131–133, 139, 209–210
Manhattan Project, 72
Margutti, Raffaella, 83
Marrone, Dan, 200–204
Marshall Space Flight Center, 108

mass: chirp mass, 160; of neutron stars, 38, 71–72, 134, 143–144, 148–155, 158–161; rotational velocity and, 154–155; of white dwarfs, 65–67
MASTER, 28
Maxie, 103
McIver, Jess, 26
McLaughlin, Maura, 234, 236, 239
McLerran, Larry, 145
MeerKAT: design and construction of, 93, 124, 148–152, 241; Galactic Center photographed by, 163–164; MeerTIME project, 153–154, 212–213; Square Kilometre Array dishes at, 244; testing of general relativity with, 228
MeerTIME project, 153–154, 212–213
Mercury, advance of periastron of, 229–230
mergers: of black holes, 205–207; of galaxies, 199, 205–206. *See also* neutron star mergers
Messick, Cody, 24–25
Messier 87 (M87), 202–206
metals, formation of, 11, 31–33, 87–88, 241
Metzger, Brian, 20, 28–29, 34
Michaelis, Anthony, 50
microwave haze, 171–172
Migdal, Arkady, 134–135
Milagro detector, Galactic Plane observed with, 167–168
Milky Way: "black widow" pulsars in, 104; collision course with Andromeda, 205; low mass x-ray binaries in, 101; magnetars in, 126; number of neutron stars in, 10; supernovas in, 47, 76. *See also* Galactic Center; Sagittarius A*
millisecond pulsars, 99–105; definition of, 4; in Galactic Center, 180–184; MSP J0740+6620, 210; planets orbiting, 188; PSR B1937+21, 154; triple systems, 222–228
Mills, Bernie, 75
Milner, Yuri, 237–238, 240
modified Newtonian dynamics (MOND), 216
Molonglo Observatory Synthesis Telescope (MOST), 73–75, 245
moment of inertia, 133
MOND (modified Newtonian dynamics), 216
Monogem ring pulsar, 170

Moon: Earth–Moon–Earth communication, 92; Earth–Moon–Sun system, 225–226; size of, 9; surface temperature measurement of, 179
moonbounce, 92
MOST (Molonglo Observatory Synthesis Telescope), 73–75, 245
Moyer, Michael, 2
MSP J0740+6620, 210
multibeam receivers, 126–128
multi-messenger astronomy, 2, 7–8, 19–22
Musk, Elon, 198

N49, 108
Nançay Radio Telescope, 210
NANOGrav (North American Nanohertz Observatory for Gravitational Waves), 210
Narkevic, David, 232–234, 236
NASA: Chandra X-ray Observatory, 101; Deep Space Network, 55, 199; Einstein X-ray Observatory, 116–117; Fermi Gamma-Ray Space Telescope, 85, 171–172; Helios 2 spacecraft, 107; Marshall Space Flight Center, 108; Voyager spacecraft, 54, 58–59
National Radio Astronomy Observatory (NRAO), 83, 178–179, 206
National Radio Quiet Zone, 177
nebulas, 73, 81–82
neon, formation of, 87
Netherlands Institute for Radio Astronomy (ASTRON), 92–93. *See also* LOFAR (Low Frequency Array)
neutralinos, 169
neutrinos: creation of, 85, 87, 138; detection and observation of, 7–8, 22, 77–78
neutron cores, Landau's theory of, 38, 70–71
neutron drip point, 138
neutrons: atomic theory, development of, 35–37, 67–68; collective quantum behavior of, 135, 137–138; discovery of, 36, 37, 67–68; formation of, 32; in neutron star core, 142
neutron star formation. *See* formation of neutron stars
Neutron Star Interior Composition Explorer (NICER), 103, 155–158, 199

neutron star mergers: accretion disks resulting from, 11; detection and observation of, 4–8, 23–31; gravitational waves and, 215–222; kilonovas resulting from, 20, 28–29, 33–34
neutron star structure: cooling and, 137; crust, 137–138; density, 9, 11, 138–139; equation of state, 142–144, 161–162; habitable zones around, 197; inner core, 141–148; magnetic fields, 57, 97–98, 169; "nuclear pasta," 139; outer core, 139–141; quarks, 141–142; radii, 153–154, 160–162; size, 9; superfluidity, 110–111, 135–141; upper mass limit, 38, 71–72, 134, 143–144, 148–155, 158–161. *See also* discovery of neutron stars; radiation emission; radio wave emission; rotational velocity
Newton, Isaac, 12, 214
Nice, David, 209
NICER (Neutron Star Interior Composition Explorer), 103, 155–158, 199
nickel, origin of, 32–33
Nissanke, Samaya, 23, 160
North American Nanohertz Observatory for Gravitational Waves (NANOGrav), 210
novas, definition of, 68
NRAO (National Radio Astronomy Observatory), 83, 178–179, 206
nuclear fusion, 71, 87
"nuclear pasta," 139
Nuclear Science Advisory Committee, 146
nucleons, collective quantum behavior of, 135
Nuijens, Frank, 92, 93, 95
NuSTAR, 83, 85, 125

"On Dense Matter" (Fowler), 66
1E 2259+586, 117, 119–120
Onsager, Lars, 136–137
Ophiuchus, hydrogen superbubble in, 180
Oppenheimer, Robert, 38–39, 71–72, 134
Orion, 80, 180
outer core, composition of, 139–141
Owens Valley Radio Observatory (OVRO), 244, 253
oxygen, formation of, 87

Pacini, Franco, 50, 76, 97
Packard, Richard, 137

PALFA (Pulsar Arecibo L-band Feed Array), fast radio bursts observed by, 238–242
Palomar Transient Factory, 79
PAMELA (Payload for Antimatter Matter Exploration and Light-Nuclei Astrophysics), 169–170
Paranal Observatory, 28, 61
Parkes Multibeam Pulsar Survey of the Magellanic Clouds, 233
Parkes Multibeam Survey of the Galactic Plane, 126
Parkes Observatory: Breakthrough Listen influence with, 243–244; design of, 93; limitations of, 244; magnetars observed with, 124–125; multibeam receiver at, 126–128; neutron star mass calculations by, 153; Parkes Pulsar Timing Array, 209; pulsar detection and observation with, 51–56; pulsar timing with, 130; Vela glitch observed with, 132–133; x-ray magnetar observed with, 121–122
Parkes Pulsar Timing Array (PPTA), 209
Parsons, William, 75
Pauli exclusion principle, 65
Payload for Antimatter Matter Exploration and Light-Nuclei Astrophysics (PAMELA), 168–169
Penzias, Arno, 36
periastron, advance of, 229–230
Perley, Dan, 79
Pethick, Chris, 134, 136, 145
phased-array feeds, 252
Pines, David, 134, 136
Pioneer Venus Orbiter, 107
planetary orbits around pulsars: Lyne's retracted paper on, 185–188; Wolszczan's discovery of, 188, 192–197
planet formation, 197
platinum, origin of, 11, 31–33
Pleunis, Ziggy, 248
polarization of radio waves, 132–133
positron excess, 167–171
post-Keplerian parameters, 228–230
p-p-dot diagram, 91
PPTA (Parkes Pulsar Timing Array), 209
precession, geodetic, 219
Prince, Tom, 118

Prognoz 7 satellite, 107
Project Ozma, 179
protons: atomic theory, development of, 35–37, 67–68; collective quantum behavior of, 135, 137–138; in core-collapse, 88, 95; formation of, 32
PSR 1913+16, 13–14, 152
PSR B0529–66, 55
PSR B1829–10, 186–188
PSR B1937+21, 154
PSR J0250+5854, 90–92
PSR J0737–3039A/B system. *See* Double Pulsar
PSR J0740+6620, 143
PSR J0952–0607, 99–100
PSR J1023+0038, 104
PSR J1614–2230, 143
PSR J1622–4950, 124
PSR J1745–2900, 125
PSR J1748–2446ad, 154
PTA (European Pulsar Timing Array), 210
Pulsar Arecibo L-band Feed Array (PALFA), fast radio bursts observed by, 238–242
pulsar kicks, 84–87
Pulsars (Manchester and Taylor), 132
pulsars: accretion-powered, 99–105; Anomalous X-ray Pulsars, 117–121; Bell's discovery of, 1–2, 44–49, 72–73; binding energy of, 226; "black widow," 104; Crab Nebula, 51, 70, 75–76, 192; Crab Pulsar, 51, 85, 98, 233; definition of, 1; in Galactic Center, 180–182; gamma-ray excess and, 182–184; Gold's theory of, 50–51; interstellar medium and, 30, 40–41, 56–59; magnetic fields of, 97, 169; measuring mass of, 148–155; moment of inertia in, 133; numbers of, 10; orbit calculations for, 218; origin of name, 50; Pacini's model of, 50; planets orbiting, 185–188, 192–197; positron excess produced by, 167–171; post-Keplerian parameters of, 228–230; precision of, 57; pulsar kicks, 84–87; pulsar timing arrays, 118, 207–213; radio waves emitted by, 4, 9; rotational velocity of, 95–96; rotation-powered, 96–99; RRATs (rotating radio transients), 233–234; testing general relativity with, 217–222; triple systems, 222–228; x-ray, 116–117. *See also* "glitches"; millisecond pulsars; neutron star structure; pulsar timing; *individual pulsar names*
pulsar timing: concept of, 129–130; as GPS beacons, 198–200; Kouveliotou's research on, 113–118; measuring pulsar mass with, 148–155; pulsar timing arrays, 118, 207–213; testing general relativity with, 217–222; three-second periodicity, 220–221; triple systems, 222–228
pulsar timing arrays, 118, 207–213

QCD (quantum chromodynamics), 142
quantum mechanics: general theory of relativity and, 214–215; superfluidity, 110–111, 135–136
quantum vortices, 136–137
quarks: discovery of, 141–142; in neutron star core, 141–148; quark-gluon plasma, 145, 147
quasar 3c279, 204
quasars: Bell's observations of, 43–44; Schmidt's observations of, 36, 40; scintillation of, 40–41
quasi-stellar radio sources. *See* quasars
Quintero, Luis, 190

Radhakrishnan, Venkatraman, 132–133
radiation emission: accretion-powered, 99–105; cosmic microwave background, 36, 200; dispersion of, 58; electromagnetic, 9, 22, 39, 171, 178; excess in, 171–177; of magnetars, 105–118; rotation-powered, 96–99; Shapiro delay, 130, 218; synchrotron, 169; timing of, 113–118, 129–130. *See also* gamma rays
radii of neutron stars, measurement of, 153–154, 160–162
radioactive decay chains, 37
radio galaxies, 41
radio pulsars, 1, 4, 126–128
Radio Schmidt telescope, 151
radio source scintillation, measurement of, 40–49
radio wave emission: from AT2018cow ("the Cow"), 82–84; detection and observation of, 4, 9, 30, 39–49; low frequency, 94–95; by magnetars, 121–124; polarization of, 132–133; scintillation of, 40–49, 58; wavelength of, 33
Raman, C. V., 64
Ransom, Scott, 206–211
Rapid Eye Mount Telescope, 26
rapid neutron-capture process, 32

Ravenhall, Geoff, 134, 136
Read, Jocelyn, 158–159
recycled millisecond pulsars, 99–105
recycling process, 103–104
red giants, accretion and, 100–101
red supergiants, 77–82, 87–89
Rees, Martin, 237
Reifenstein, Edward, 75
Relativistic Heavy Ion Collider (RHIC), 146–147
"relativistic" neutron star systems, 230
relativity, theory of. *See* general theory of relativity
RHIC (Relativistic Heavy Ion Collider), 146–147
Robert C. Byrd Green Bank Telescope. *See* Green Bank Telescope (GBT)
ROSAT, 31
Rossi X-ray Timing Explorer (RXTE) satellite, 113–114, 118
rotating radio transients (RRATs), 233–234
rotational velocity: of Anomalous X-ray Pulsars, 118–120; magnetic field and, 111–113; mass and, 154–155; of pulsars, 86, 95–96; superfluidity and, 139
rotation-powered pulsars, 96–99. *See also* LGM-1 pulsar; PSR J0250+5854
Royal Astronomical Society, 49, 50, 67
r-process, 32
RRATs (rotating radio transients), 233–234
Rubin, Vera, 165
Ruderman, Mal, 136
Rutherford, Ernest, 35, 37, 68, 135
RXTE (Rossi X-ray Timing Explorer) satellite, 113–114, 118
Ryle, Martin, 40, 41, 47, 49

Sagan, Carl, 255
Sagittarius A* (Sgr A*), 150, 164, 172, 203
Sanduleak −69° 202, 77–78
San Pedro de Atacama, 60
SARAO (South African Radio Astronomy Observatory), 121
Sardinia Radio Telescope, 210
Sarkissian, John, 54–58, 122, 127, 141
Sazhin, Mikhail, 207–208
Schmidt, Maarten, 36, 40
Scholz, Paul, 238, 239, 241, 245
Schwarzschild, Karl, 204

Schwarzschild radius, 204
scintillation, measurement of, 40–49, 58
Scorpius X-1 (Sco X-1), 102
Scott, Dave, 226
Scott, Paul, 46
"scruffs," Bell's discovery of, 36, 44–49
Scutum, 186
SDSS J0849+1114, 205
Search for Extraterrestrial Intelligence (SETI), 179, 235–236, 254
SEXTANT (Station Explorer for X-ray Timing and Navigation Technology), 199
SGR 1627−41, 116
SGR 1806−20, 242
SGR 1900+14, 116
SGRBs. *See* short gamma-ray bursts (SGRBs)
SGRs (soft gamma repeaters), 105–110, 114–116
Shannon, Ryan, 249–254
Shapiro delay, 130, 218, 221, 228–229
Shelton, Ian, 77
Shemar, Setnam, 186–188
Shklovsky, Iosif, 102, 134
short gamma-ray bursts (SGRBs): emission by neutron stars, 20; observation and detection of, 29–30
silicon, formation of, 87
silver, origin of, 11, 31–33
Sirius, 65
Sirius B, 65
SKA (Square Kilometer Array), 148, 151–152, 227–228, 244, 256
Slatyer, Tracy, 171–177, 181–184
slit spectroscopy, 79
Small Astronomy Satellite-1 (SAS-1), 102
Small Magellanic Cloud (SMC), 55, 129, 234
Smartt, Stephen, 21, 63, 78
SN 185, 69
SN 1054, 70
SN 1987A, 77–78, 82
SN 2013fs, 78–81
SN 2018gep, 81
soft gamma repeaters (SGRs), 105–110, 114–116
sound waves, as warning sign of supernovas, 81
South African Radio Astronomy Observatory (SARAO), 121

South Pole Telescope (SPT), 200–204
spacetime, 10, 13, 228
Special Breakthrough Prize in Fundamental Physics, 49
Spectroscopic Time-Resolving Observatory for Broadband Energy X-rays (STROBE-X), 157–158
Spiewak, Renee, 52
spin-down rate, 113–119
Spitler, Laura, 238, 249
Spitler Burst, 238
SPT (South Pole Telescope), 200–204
Square Kilometer Array (SKA), 148, 151–152, 227–228, 244, 256
Staelin, David, 75
Stairs, Ingrid, 209
Stanford Linear Accelerator Center, 141
starquakes, 112. *See also* magnetars
Station Explorer for X-ray Timing and Navigation Technology (SEXTANT), 199
stellar forensics, 76–78
stellar-mass black holes, 200–207
Stevin, Simon, 226
strange quarks, 142
STROBE-X (Spectroscopic Time-Resolving Observatory for Broadband Energy X-rays), 157–158
Su, Meng, 171–172
Sun: Earth–Moon–Sun system, 225–226; magnetic field of, 111; x-ray emission from, 101–102
superattenuators, 16
Superconducting Super Collider, 146
superfluidity: dynamo effect and, 110–111; in neutron star core, 138–141; process of, 135–136
supergiants, 4, 77–82, 87–89
superluminous supernovas, 89
supermassive black holes, 203–205
supernovas: AT2018cow ("the Cow"), 63–64, 82–84; circumstellar material in, 79–80; core-collapse, 77–82, 87–89; definition of, 3–4; detection and observation of, 7; explosion of, 3–4, 9–10; historical observation of, 69–70; kilonovas, 11; nebular phase of, 81–82; neutrino-driven mechanism of, 7–8, 77–78, 85; prevalence of, 76; stellar forensics of, 76–78; Type-I superluminous, 241; warning signs of, 76–82; Zwicky's theory of, 38, 67–73. *See also individual supernova names*
Superterp, 95
Swank, Jean, 119
Swift telescope, 103, 125
Swope Supernova Survey 2017a (SSS17a), 26
synchrotron radiation, 169

Tan, Chia Min, 90–92
Tarantula Nebula, 77
Tau Ceti, 179
Taurus, SN 1054 in, 70
Taylor, Joseph Hooton, Jr., 13, 152, 192, 206, 219, 229
tensor-vector-scalar (TeVeS), 216
tera-electron-volts (TeV), 167
TeV (tera-electron-volts), 167
TeVeS (tensor-vector-scalar), 216
theory of relativity. *See* general theory of relativity
theory of universal gravitation, 12
Thompson, Chris, 110–117, 126
Thomson, J. J., 36
Thorne, Kip, 6, 14, 19
Thornton, Dan, 236–237, 245
Thousand Pulsar Timing Array (TIME), 212
three-second periodicity, 220–221
tidal effects, 10–11
TIME (Thousand Pulsar Timing Array), 212
timing, magnetar, 118–120, 245
timing, pulsar: concept of, 129–130; as GPS beacons, 198–200; Kouveliotou's research on, 113–118; mass measurement with, 148–155; measuring pulsar mass with, 148–155; pulsar timing arrays, 118, 207–213; scientific uses of, 198–200; testing general relativity with, 217–222; three-second periodicity, 220–221; triple systems, 222–228
timing arrays, solar system as, 207–213
Tolman, Richard Chace, 71
Tolman-Oppenheimer-Volkoff limit, 71–72
tool, pulsars used as: GPS beacons, 198–200; gravitational wave detection, 200–207; Keplerian and post-Keplerian parameters,

228–230; pulsar timing arrays, 207–213; tests of general relativity, 213–222; triple system, 222–228
Townsend, Lee, 129
transitional millisecond pulsars, 104–105
triple system, discovery of, 222–228
Type Ib core-collapse supernovas, 81, 88–89
Type Ic core-collapse supernovas, 88
Type II core-collapse supernovas, 88–89
Type-I superluminous supernovas, 241

Uhuru satellite, 102
ultra-luminous x-ray sources (ULXs), 128–129
universal gravitation, theory of, 12
Universe, expansion of, 32–33
University of Cambridge, Cavendish Laboratory at, 35–37, 135. *See also* Interplanetary Scintillation Array
upper mass limit: of neutron stars, 38, 71–72, 134, 143–144, 148–155, 158–161; of white dwarfs, 65–67
URSI (International Union of Radio Science), 151

Vanderline, Keith, 246
Vaughan, Alan, 74–75
Vela Pulsar, 51, 75, 85, 131–133, 141
Vela satellites, 29, 105
velocity. *See* rotational velocity
Venera 11/12 spacecraft, 106–107
Venus, surface temperature measurement of, 179
Verlinde, Erik, 217
Very Large Array, 30, 83, 195, 239–241
Very Large Telescope (VLT), 61
Victor M. Blanco Telescope, 28
Virgo interferometer: Advanced Virgo upgrades to, 18–19; black hole merger detected by, 22–23; funding and construction of, 14–18; GW170817 merger detected by, 6–7, 10–12, 23–31, 159–161; limitations of, 206–207; management of, 15; in multi-messenger astronomy project, 19–22
Visible and Infrared Survey Telescope for Astronomy (VISTA), 28
VLT (Very Large Telescope), 61

Volkoff, George, 38, 71–72, 134
vortices, quantum, 136–137
Voyager spacecraft, 54, 58–59
Vulpecula, 47

Wajarri Yamatji people, 241
Watts, Aeryn, 155
Watts, Anna, 155–158
wavelength, color and, 33–34
weakly interacting massive particles (WIMPs), 169, 171, 173, 177, 183
Weber, Joe, 207
Weisberg, Joel, 219
Weiss, Rainer, 6, 14, 19
Weniger, Christoph, 181
Westerbork Telescope, 104, 210
white dwarfs: Chandrasekhar's research on, 65–67; cooling of, 65, 87; density of, 65–67; formation of, 104; historical focus on, 38; planets orbiting, 197; rotation of, 49–50; upper mass limit of, 65–67
Wilkinson Microwave Anisotropy Probe (WMAP), 171
Wilson, Robert, 36
WIMPs (weakly interacting massive particles), 169, 171, 173, 177, 183
WMAP (Wilkinson Microwave Anisotropy Probe), 171
Wolszczan, Alex, 188–197, 222, 230
Woods, Pete, 119

X-DINS (X-ray Dim Isolated Neutron Stars), 96
xeXTP (enhanced X-ray Timing and Polarimetry), 157
XMM, 31, 125
XMM-Newton, 101
X-ray Dim Isolated Neutron Stars (X-DINs), 96
X-ray Einstein Observatory, 107
x-rays: Einstein X-ray Observatory, 116; high mass x-ray binaries, 103; from magnetars, 105–106, 123–124; from neutron stars, 96–97; Rossi X-ray Timing Explorer satellite, 113–114, 118; Spectroscopic Time-Resolving Observatory for Broadband Energy X-rays, 157–158; from Sun, 101–102; ultra-luminous x-ray

x-rays (*continued*)
sources, 128–129; x-ray pulsars, 116–117.
See also radiation emission
XTE J1810 – 197, 121–125

Yaron, Ofer, 79–80

Zumalacárregui, Miguel, 213–215, 227
Zwicky, Fritz: *Catalogue of Galaxies and of Clusters of Galaxies*, 73; dark matter theory of, 72, 164–165, 214; death of, 73; supernova theory of, 37–38, 50, 67–73
Zwicky Transient Facility, 81